THE CABLE

THE CABLE

Wire to the New World

GILLIAN COOKSON

The
History
Press

To my friend
John Hirst Fenton
1928–1994

This new edition has been produced in conjunction
with the Radio Society of Great Britain.

First published 2003
This edition published 2012

The History Press
The Mill, Brimscombe Port
Stroud, Gloucestershire, GL5 2QG
www.thehistorypress.co.uk

British Library Cataloguing in Publication Data.
A catalogue record for this book is available from the British Library.

ISBN 978 0 7524 8786 1

Typesetting and origination by The History Press
Printed in Great Britain

Contents

Acknowledgements

Thanks especially to Colin Hempstead, and for information, help, advice, support and encouragement variously to Bill Burns of the Atlantic Cable website (http://www.atlantic-cable.com/); Charlotte Dando, collections manager, Porthcurno Telegraph Museum, and Allan Green, research fellow at the museum; Anne Locker, archivist of the Institution of Engineering and Technology (formerly the Institution of Electrical Engineers); Amy Rigg and Abigail Wood of The History Press; and to Neil, Joe and Francis Cookson.

1

The Mystic Voice
of Electricity

Whose idea was the transatlantic cable? Once the scheme was a success, and even before that, there was no shortage of claimants. John Watkins Brett declared in 1857, as the first Atlantic cable was being manufactured, that he and his brother Jacob had thought of it early in 1845. By then, a network of land telegraphs had spread quickly across England, following Wheatstone and Cooke's 1837 patent. Most of these were overhead lines, but some were underground. 'If possible underground, why not under water?' asked the Brett brothers. And if under water, could it not lie on the bed of the ocean?

The Bretts were convinced that the idea would work, and went so far as to register a company with the aim of establishing a telegraph from Britain to Nova Scotia and Canada. They also tried to interest the British government in an experiment, a cable through which Ireland could be ruled by putting 'Dublin Castle in instantaneous communication with Downing Street'. This came to nothing.

But the Bretts were much more than dreamers. They made their name in submarine telegraphs, laying the first international line beneath the sea, between France and England in 1851. In 1845, though, their idea had barely materialised into any real shape. At that time they could not conceivably have crossed the Channel, let alone the Atlantic. Years later, John Brett discovered that Wheatstone himself had also in 1845 been developing a scheme for a Channel telegraph. Yet Brett continued to insist that submarine cables were 'purely an invention of our own' and that 'no man's labours or suggestions were borrowed'.

In the United States, Professor Samuel Finley Breeze Morse would have disputed this point. Morse was some years ahead of the Britons. Cromwell Fleetwood Varley, an eminent British telegraph engineer, later had no doubt that the transatlantic cable had originated in America: 'It is indisputably clear that the idea of connecting the US with England practically originated in New York, that these American originators pushed on the telegraph.' The word 'practically' is key. It was simple to visualise a long submarine line, but few were capable of advancing the idea any further. Morse proved that he was one of that small number by laying a cable in New York harbour, across part of the East River, in 1842. Although an anchor destroyed the line almost as soon as it was operational, he had shown beyond doubt that the feat was possible.

Samuel Morse, who had started out as an artist, was professor of Natural Science at Yale and a pioneer of land telegraphs in the United States. His experiments on submarine cables were well recorded, so his claim that he was thinking about a transatlantic cable early in the 1840s is convincing. Morse left an account of how his ideas at that time had developed, in letters written during 1854 to General Horatio Hubbell of Philadelphia. Hubbell's role in the ocean telegraph is insignificant, except that he too was claiming first inspiration for the idea. On the strength of this, he wrote to Morse demanding a seat on the board of a company with which the professor was involved. Morse replied at some length, explaining the evolution of his own work with perhaps more patience than Hubbell merited:

It was quite natural that the extension of my system throughout the world should occupy my thoughts with some degree of intensity, and that in view of this anticipated world-wide extension, the connection of Europe and America was at least a possible, if not a probable subject of thought and speculation with me. Now this is a subject which occupied my mind at least as early as 1842, as printed documents before Congress elucidate.

The idea then was 'a brilliant but impracticable, or rather unsolved, conception', as unreal, said Morse, as air travel. He tried to point out to Hubbell that the notion itself amounted to little:

A claim for the original barren thought, however brilliant, is comparatively of little account in the eyes of the world. It is he who first combines facts, plans and means to carry out a brilliant thought to a successful result who in the judgment of the world is most likely to receive the greatest credit, while, nevertheless, an impartial posterity will award to each one whose mind has been employed in elaborating any part of a useful project his just share of honor in bringing it to a result.

And Morse went on at length to list his own thought processes as the idea had taken hold of him. In 1842 there had been many great unknowns:

First, can electricity, by means of a single electro-motor, be propelled to a distance so great as the width of the ocean? This was a problem which my experiments of 1842, 1843, were intended to solve and which was so far satisfactorily solved to my own mind, as to lead me to declare the law of propulsion, or rather the law of battery construction.

As it turned out, this conclusion was optimistic and very premature. But Morse could certainly prove that he had been working on the question then, for he described the experiments to the secretary of the US Treasury in August 1843. His report to the Treasury concluded with the words: 'The practical inference from this law is that a telegraphic communication on my plan may with certainty be established across the Atlantic. Startling as this may now seem, the time will come when this project will be realized.'

Morse's second problem was information about the state of the ocean bed. 'This bed had not then been sounded, and, therefore, its character, whether suitable or not … for the reception of a proper conductor, was not known.' The United States Navy had since taken ocean soundings, especially for the purpose of laying a cable, but these were still, in 1854, far from complete, so that this question was not fully resolved.

The third problem, continued Morse, was 'can a cable conductor of such a length be paid out to such a depth as is required?'

> This is resolved only by conjecture, and by the experience of comparatively very short distances in successful submarine crossings of rivers and wide channels. The first attempt for telegraphic purposes was made, so far as I believe, by me across the East River between Castle Garden and Governor's Island in the autumn of 1842. Long subsequent to this submarine experiment, English companies have laid the conductors between the Irish and the English channels.

Morse could have gone into much more detail about the host of mechanical and electrical unknowns. These questions could be resolved only by laying a long cable, and observing how well the process worked. How would the cable be constructed, protected, handled, laid? Would new kinds of instruments and electrical testing, and refined systems of working, be required? In these, as with the great electrical question of whether a signal could pass the width of the Atlantic Ocean, experiments in a laboratory or on a small scale were of limited use. The only way to know was to try.

The fourth problem was finance. Hubbell's efforts to lobby government for funding during the past five years had met responses on the lines that 'the world is not yet ready for such a project'. Morse knew that such an enterprise could go nowhere without massive backing from government or private enterprise. 'Means are as essential to the project before it can be made of practical value, as all other parts of it.' Although Morse was by then involved in such a company, and had been key to persuading it to broaden its horizons and to consider the Atlantic crossing, he knew there was a long way to go.

It was the formation of this company – the New York, Newfoundland & London Electric Telegraph Co. – which had spurred Hubbell to approach Morse. As originator of the whole idea, he considered himself entitled to a seat on the board. Morse tolerantly explained that he had misunderstood the role of directors, and that the company was open only to 'men of great capital'.

Hubbell ignored the suggestion that he become a director by making a large investment. The General, undeterred by Morse's scientific eminence, was more interested in pressing his technical views. When he proposed that the Atlantic cable should be buoyed rather than laid on the ocean floor, Morse's patience ran out. Such a system made a cable vulnerable to 'disturbing and disastrous agencies of storms, currents, ice and malevolence'. Morse made it clear that he was not prepared to consider Hubbell's ideas further. General Hubbell took offence, writing a 'discourteous and acrimonious rejoinder', and was thereafter a small but irritating thorn in the side of the Atlantic telegraph, using patents to try fruitlessly to checkmate the project.

The new company which so annoyed Hubbell had evolved from a less ambitious scheme in 1850. Although the Brett brothers failed in their first attempt to lay a cable across the Channel that year, a belief was growing that underwater telegraphs were a viable proposition. This planted an idea among residents of Newfoundland, the remote and inhospitable British colony at the north-eastern point of North America. In November 1850, the Roman Catholic Bishop Mullock of Newfoundland wrote to an American newspaper suggesting that a telegraph line should be built between Newfoundland and New York. This went rather further than what Morse would have called an 'original barren thought', for it proposed to use technology which was then becoming serviceable, on the threshold of practicability. Mullock's scheme was for a land telegraph connecting St John's, the island's capital, with Cape Ray on the far west of the island, and submarine lines from there to the American mainland via St Paul's and Cape Breton islands. For St John's, it promised a significant new role as a communications centre. Passing mail steamers could be intercepted for messages from Europe to send on to New York, or might be handed the latest telegraphed news and cables from New York as they set sail to Ireland. Thus the time taken for mail to pass between Britain and the United States would be reduced by several days.

About the time of the bishop's letter, and maybe as a result of it, a similar idea came to Frederic Newton Gisborne. Gisborne was to initiate a chain of events which eventually, after many setbacks and obstacles, would lead to the first Atlantic cable. At the time when he first thought of a transatlantic line, which he later claimed happened in 1849, he was a telegraph engineer on overland cables. Born in 1824 in England, he had trained under a pupil of Samuel Morse, and by 1850 was superintendent of the Nova Scotia government telegraph lines, then the only wires in the province. Gisborne put forward a plan for a submarine cable between Newfoundland and the North American mainland at Halifax. His employers gave him leave to find support for this idea in St John's. To the amazement of the Nova Scotia commissioners, Gisborne returned from Newfoundland with a plan which had gone far beyond the Newfoundland–Nova Scotia line, to encompass a transatlantic

The north-eastern United States and
British provinces of North America.

link to Ireland. Later there was a dispute about whether he had actually proposed an Atlantic cable, or whether his idea was merely to use fast steamers across the ocean. At that time, an Atlantic cable cannot have been much more than a theoretical prospect. But whether or not he put the transatlantic cable to them, the commissioners would not allow Gisborne to raise capital for the scheme, and he parted company with them in the summer of 1851 when the Nova Scotia telegraphs were transferred to a private company. He left for St John's to pursue his transatlantic vision.

Gisborne's first challenge was to find a route along the south coast of Newfoundland from St John's to Cape Ray, on the far west of the island and the closest point to Nova Scotia. Between the two places, 300 miles as the crow flies, was *terra incognita*, unmapped and inhospitable wilderness impenetrable even on horseback. The forest was dense, the terrain marked with rocks and marshes, the climate foggy. Local fauna included bears and wolves. But Gisborne had a reputation for physical toughness. During a previous expedition in Quebec, he had once snow-shoed 100 miles while dragging all his possessions on a toboggan, for which his employer had presented him with a medal. He was to become known as 'the indomitable electrician'. The Newfoundland survey, though, was exceptionally arduous even by his standards. The engineering difficulties, the personal dangers and privations in this unexplored territory were formidable. The six men working for him deserted after the first hundred miles. He recruited instead four native Americans, one of whom died after a few days. Two others fled, and the fourth, who lasted to the end, never afterwards recovered his health.

Yet Gisborne was back in St John's in December, only three months after setting out on this hazardous and exhausting mission, and soon afterwards left for Boston in the hopes of finding support there. New York proved to be a more promising source of funds, and from there he went to London to meet John Brett. When he had set out to survey the line to Cape Ray, Gisborne expected that messages might have to cross the sea to Nova Scotia by carrier pigeon or steamship. But Brett's success with the Dover-Calais line helped crystallise what had up to then been only an abstract idea. Gisborne bought a cable made by R.S. Newall and the Gutta Percha Co. in England, and very quickly organised the necessary finance, technology and machinery, much of it of his own design. He set up the Newfoundland Electric Telegraph Co., using capital supplied by Brett, and by two New York businessmen, Horace B. Tebbets and Darius B. Holbrooke, and secured telegraph rights on the island for thirty years:

> This company is designed to be strictly an inter-continental telegraph. Its termini will be New York and London; these points are to be connected by a line of electric telegraph from New York to St John's, partly on poles, partly laid in the ground, and partly through the water, and a line of the swiftest steamships ever built from that point to Ireland. The trips of these steamships, it is expected, will not exceed five days, and as very little time will be occupied in transmitting messages between St John's and New York, the communication between the latter city and London or Liverpool will be effected in six days, or less. The company will have likewise stationed at St John's a steam yacht, for the purpose of intercepting the European

and American steamships, so that no opportunity may be lost in forwarding intelligence in advance of the ordinary channels of communication.

Compared with the scheme which followed soon afterwards, this first plan of Gisborne's seems unambitious. Yet the goal of a six-day crossing for news and messages had still not been achieved more than a decade later, when twelve days was the best that could be managed, even for the most urgent information.

Relations with the Nova Scotia Electric Telegraph Co. had soured further. They demanded charges which Gisborne considered exorbitant to handle messages on their lines, so the route from Cape Ray was changed, to pass further north across the Cabot Strait via Prince Edward Island, reaching

The *Goliah* laying the first Dover–Calais cable, 1850.
(*Illustrated London News*)

the mainland at New Brunswick. Gisborne's cable between Prince Edward Island and New Brunswick, a distance of twelve nautical miles and to a depth of fourteen fathoms, laid by November 1852, was the first submarine line in America.

This cable was soon afterwards broken, and Gisborne had neither the technology nor the experience to retrieve and repair it. Instead he started in earnest to raise capital for the transatlantic crossing. He worked in New York with Tebbets and Holbrooke, while Brett undertook fund-raising in Britain. Gisborne had also to supervise the construction of his line across Newfoundland, a scheme which involved building a road, 'a good and traversable bridle-road, eight feet wide, with bridges of the same width', according to the agreement. A force of 350 labourers was employed on this vast project. Gisborne intended to lay the wires underground, but in the spring the ground was frozen too hard even to dig post-holes. The only possibility was to support poles within stone embankments made of rocks piled on the surface. Forty miles of line had been built when Tebbets and Holbrooke quarrelled over finances and refused to make any further payment to Gisborne. The engineer narrowly escaped imprisonment for fraud, lost everything he owned and had debts of over $50,000, most of it owed in wages to the workmen on the line.

Apparently inexhaustible, Gisborne returned to his quest for funding. Brett agreed to carry on in Britain, suggesting to Gisborne that he offer the Newfoundland government free transmission of their messages if they would set up a new telegraph company and endow it with exclusive landing rights on the island for fifty years. Experience had already shown Brett that such contracts were essential to raise sufficient money for submarine cables, which had exceptionally high capital costs. Without the promise of a monopoly and high returns, investors were not willing to back such a risky project.

The young engineer set out again for New York in the hope of raising further support. There he made the acquaintance of Matthew D. Field, a civil engineer, and through him was introduced to the engineer's brother, Cyrus West Field. This chance meeting entirely altered the direction and the prospects of the Atlantic telegraph.

Cyrus Field was thirty-four years old and had already retired from business. He was then worth $250,000. Born in 1819 the son of a Congregational minister in Stockbridge, Massachusetts, from a modest start he had made a fortune, lost it, repaid his debts and acquired enough money to step down from his company when advised by doctors that he was overworking. As a boy, Field had turned down a college education, instead serving his apprenticeship at a dry goods store in New York. He learned to sell, discovering the pitfalls of commerce in 1837 when a panic brought ruin to large numbers of enterprises. He attended evening classes in book-keeping and penmanship, and after three years in the city returned to his home state to join his brother in a paper-making business. Field was engaged mainly in sales, through which he made many contacts with wholesalers in New York such as Elisha Root and Chandler White, and with other paper manufacturers in New England.

Field started a business of his own, but gave it up when invited to join Root as a partner. Unknown to him, Root was already insolvent. Within six months the business failed and Field found himself responsible for huge debts. He re-established his own firm, taking over Root's customers

Cyrus Field.
(*Institution of Engineering
and Technology*)

and contacts. He also brought in some of his acquaintances
from Massachusetts, building up a network of suppliers and
customers there. Root's liabilities he settled at 30¢ in the
dollar. His product was a better one than Root's, and despite
the generally depressed markets, he had found a good time to
move into high-quality paper, for which demand was huge,
and growing. In three years from 1846, he sold an astonishing
$1 million-worth of paper. But by 1849, the effort had taken
its toll, and Field was sent to Europe to convalesce.

It was Field's maiden trip across the Atlantic – the first
of what would be more than thirty visits to Britain – and
the start of his love affair with England and things English.
It was at the time when Brett's cross-channel cable was
being planned, and Field may have heard about it then. He
returned to New York, where his fortune had grown in
his absence, putting him among the city's top three dozen

richest men, and did further good to his reputation by settling fully all the money owing to Root's creditors. His business, left to the management of others, continued to make vast amounts of money without needing Field's active involvement. On his return from Europe, Field moved into Gramercy Park, on the southern end of Lexington Avenue. Sometimes called the American Bloomsbury, Gramercy is the nearest thing to a London square that New York can offer. At its centre is a park kept exclusively for residents. The newly-built house into which Field moved in about 1852 has not survived, but there is still a plaque commemorating the role of this neighbourhood in advancing the cause of the transatlantic cable.

Field's new neighbours were rich and influential New Yorkers – newspapermen, politicians, artists and businessmen. Next door was Peter Cooper, a self-made industrialist who had dabbled in railroads and land telegraphy, and who owned a factory in New Jersey which made telegraph wire. Cooper, then in his sixties, had much in common with Field despite the age difference, and the two became close. Another brother of Field, David Dudley Field, a lawyer, was also a neighbour. Through the social gatherings of Gramercy Park, Field and Cooper made many useful contacts, among them Samuel Morse.

For a while, Field's attention was taken up with his home. He was the first New Yorker to hire an interior decorator from France. His house was fitted out with heavy Italian drapes, Greek statues, marble and frescoes. Field recruited an English butler, who served dinner on Minton china. Mary Field, whom he had married when he was twenty, just before the Root failure, had the first private greenhouse in New York.

Her husband delighted in his library. Behind the house, in the still rural surroundings of the reclaimed swampland, the Field family kept a cow along with their horses.

This inactive life, though, did not satisfy Field. 'I never saw Cyrus so uneasy as when he was trying to sit still,' said his brother Matthew. At this point, Matthew Field, between contracts after years away in the southern and western states building railroads and bridges, fell into conversation with Frederic Gisborne in the lobby of the Astor House hotel in January 1854. Gisborne described his struggling project, and the civil engineer recognised its possibilities. He saw that his wealthy and under-occupied brother may want to be involved, and took Gisborne to Gramercy Park to explain the scheme.

Cyrus Field knew very little about telegraphy and was ignorant of the geography of the British provinces north of the United States. He was not enthused by the limited project to link Newfoundland to New York, but the idea of a transatlantic, and maybe then a global, cable gripped his imagination. From the start he saw this telegraph line as a means of deepening international understanding and harmony, especially between the two countries he loved, the United States and Britain. Of course, commercial life would also benefit from improved communications. Not least, investors in the cable stood to make a great deal of money.

It was disputed later who within the company had first seriously proposed spanning the Atlantic with a cable. Gisborne certainly knew at the time the English Channel was crossed successfully by a telegraph line in 1851 that this new technology had the potential to link Newfoundland to the North American continent. Later he hinted that an Atlantic cable had also been on his agenda, explaining that he had not publicised the idea:

> I was looked upon as a wild visionary by my friends, and pronounced a fool by my relatives for resigning a lucrative government appointment in favor of such a laborious speculation as the Newfoundland connection. Now had I coupled it at that time with an Atlantic line, all confidence in the prior undertaking would have been destroyed, and my object defeated.

Field's admirers, though, disputed whether Gisborne's transatlantic plan had ever gone further than a vague notion that such a thing may be possible at a distant time in the future.

But by 1854 the idea of a cable across the ocean had already progressed far beyond an abstraction. Immediately after his meeting with Gisborne, Field started to take the best technical advice that was available. He wrote to the oceanographer Lieutenant Matthew Fountaine Maury, head of the Naval Observatory in Washington, and also to Samuel Morse. He already knew Maury slightly, having corresponded with him the previous year when planning a trip to South America. When Field's letter arrived, Maury was at that very moment analysing results from a survey of the ocean bed taken in 1853 by Lieutenant Ottway H. Berryman of the US brig *Dolphin*. It is clear from Maury's reply to Field that the US government was already thinking on the same lines of a transatlantic cable.

Deep-sea soundings were still a crude affair. A cannon ball was dropped on the end of a long line, an unreliable way to measure depth as currents carried the twine away from the vertical. There had been slight improvements to this basic system, but most of the Atlantic Ocean remained unsurveyed. It was known, though, that in many places the seabed was broken and irregular, with sharp peaks and crested ridges. This could be fatal to a cable. Any direct route between Europe and the United States would also have to cover an extraordinary distance, perhaps 3,000 miles, and contend with extreme depths of ocean. A relay point in the Azores would break the length, but that meant that the cable would suffer great physical risks across the western reaches of the Atlantic. There were also dangers from volcanic activity on the seabed in the approaches to the Azores. Potential routes to the north of Newfoundland were quickly discounted because of the stormy seas and dangers of floating ice. In any case the seabed there was unsuitable, and landline connections difficult because of the climate and terrain.

The shortest direct route also turned out to be the best in terms of the depth and the composition of the seabed. The US naval survey had the best possible news – the ocean bed between Newfoundland and Ireland might have been made with a cable in mind. Maury reported to Field that between the closest points of the Old and New Worlds – the west coast of Ireland and the east coast of Newfoundland – lay a distance of 1,600 nautical miles. Yet, he said:

> the bottom of the sea between the two places is a plateau, which seems to have been placed there especially for the purpose of holding the wires of a submarine telegraph, and of keeping them out of harm's way. It is neither too deep nor too shallow; yet it is so deep that the wires, but once landed, will remain for ever beyond the reach of vessels'

anchors, icebergs, and drifts of any kind, and so shallow that the wires may be readily lodged upon the bottom. The depth of this plateau is quite regular, gradually increasing from the shores of Newfoundland to the depth of from 1,500 fathoms to 2,000 fathoms as you approach the other side.

A wire laid along the shortest route would conveniently avoid the Grand Banks and rest on this 'beautiful plateau'.

Maury had sent samples from the seabed for microscopic examination at West Point. The floor of the ocean was found to be entirely made up of microscopic shells – 'not a particle of sand or gravel exists in them'. Nothing lived at these depths, but the shells had been carried from the tropics by the Gulf Stream and gradually accumulated where they dropped on to the ocean bed. Once there, they did not move, as there were no currents to disturb the depths. The surrounding water was as quiet as the bottom of a millpond. 'Consequently, a telegraphic wire once laid there, there it would remain, as completely beyond the reach of accident as it would if buried in air-tight cases.' In a few years' time, more of this harmless deposit would completely cover the cable.

Maury's discoveries were decisive in setting the course of the cable. Nature, it was said, had 'beneficently decided the question'. Newfoundland stretched forth as the hand of the New World, to meet the grasp of the Old, the British Isles:

> The course of the telegraph cable is precisely marked out by a natural tracing across the depths of the ocean. There is one line, and only one line, in which the work can be accomplished. Providence has designed that the Old

World and the New, severed at the first by a great gulf,
shall be re-connected by electrical sympathies and bonds,
and Providence has prepared the material means for the
fulfilment of the design.

Maury had written to the United States legislature to that
effect, and he christened the route 'the telegraph plateau'.
Others called it 'Maury's plateau', and the lieutenant, a
self-educated scientist, received many plaudits and honours
through his work on behalf of the cable.

While this news was very welcome, Maury had also been
contemplating some of the other difficulties that lay in the
way of the cable. He raised a question for Field's consider-
ation. It was in truth several questions: of 'the possibility of
finding a time calm enough, the sea smooth enough, a wire
long enough, a ship big enough, to lay a coil of wire 1,600
miles in length.' His tone, though, was optimistic: 'I have no
fear but that the enterprise and ingenuity of the age, whenever called on with these problems, will be ready with a
satisfactory and practical solution of them.'

Morse's vision a decade earlier, that a magnetic current could
cross the Atlantic, 'startling as this may now seem', was starting
to look less utopian. Events since the experiments in New York
harbour in 1842 had only strengthened his faith in his original judgement. Field met Morse and was further heartened.
He then set about recruiting men of wealth and probity who
would, like him, recognise the public and commercial benefits
as well as the scheme's financial promise. His brother Dudley
agreed to be the project's legal advisor. An obvious candidate
to join any scheme was Field's neighbour Peter Cooper, 'one of
the great capitalists of the New World'. Field knew that Cooper

had money to invest and some technical understanding, though the older man was preoccupied with other projects, many of them charitable. But Cooper seems to have been quickly taken by the idea, and proved to be Field's chief support through the long and arduous years ahead:

> It was an enterprise that struck me very forcibly the moment [Field] mentioned it. I thought I saw in it, if it was possible, a means through which we would communicate between the two continents, and send knowledge broadcast over all parts of the world. It seemed to me as though it were the consummation of that great prophecy, that 'knowledge shall cover the earth, as the waters cover the deep', and with that feeling I joined him in what then appeared to most men a wild and visionary scheme; a scheme that fitted those who engaged in it for an asylum where they might be taken care of as little short of lunatics. But believing as I did, that it offered the possibility of a mighty power for the good of the world, I embarked on it.

Religious vision and the promotion of peace were to be recurring themes around the Atlantic cable. The involvement of Cooper brought in other, more prosaic, characters. First was Moses Taylor, an importer and later president of the City Bank, who sat and listened to Field for an hour without saying a word. Taylor then immediately agreed to the proposal. Field and Taylor, who had not previously known each other, were to become close friends. Taylor introduced his friend Marshall O. Roberts, like Taylor a self-made man, the son of Welsh immigrants, who had made his money from railroads and a steamer line to gold-rush California via Panama. Another recruit to the project was Field's old friend Chandler White, the paper wholesaler, who had by then retired. White had moved out of the city to Fort Hamilton, a retreat with pan-oramic views of the New York harbour, and seemed to have left the cares of commerce behind him. Yet he too allowed himself to be convinced by Field's enthusiasm. Samuel Morse agreed to be the company's electrician.

This group of five or six first met at the Clarendon Hotel, where a total of $40,000 was pledged, and Cooper was nom-inated as company president, a role he would fill for two decades. Later anniversaries marked 10 March 1854 as the launch of the Atlantic project. That was the date of the first of four consecutive evenings when the company gathered in Field's dining room, the Minton porcelain moved aside to make space on the table for maps and charts. The company drew up detailed plans and costings, and speculated as to the likely profits. Every one of them agreed to join the under-taking, on condition that the government of Newfoundland improved the terms of the existing company's charter.

It was imperative to talk directly to the Newfoundland authorities. Within four days Cyrus and Dudley Field and Chandler White, accompanied by Gisborne, embarked on a journey from New York to Boston, where they took a steamer for Halifax and then a smaller ship, *Merlin*, to St John's. Dudley Field later recalled the trials of the last leg of this journey:

> Three more disagreeable days, voyagers scarcely ever passed, than we spent in that smallest of steamers. It seemed as if all the storms of winter had been reserved for the first month of spring. A frost-bound coast, an icy sea, rain, hail, snow and tempest, were the greetings of the telegraph adventurers in

their first movement towards Europe. In the darkest night, through which no man could see the ship's length, with snow filling the air and flying into the eyes of the sailors, with ice in the water and a heavy sea rolling and moaning about us, the captain felt his way round Cape Race with his lead, as the blind man feels his way with his staff, but as confidently and as safely as if the sky had been clear and the sea calm; and the light of morning dawned upon deck and mast and spar, coated with glittering ice, but floating securely between the mountains which form the gates of the harbor of St John's.

The visitors were first introduced to Edward M. Archibald, attorney-general of the colony, who warmly welcomed the scheme. He introduced them to the Governor of Newfoundland, Kerr Bailey Hamilton, who called together the island's council to meet the group and hear their plans. Newfoundland, established in 1583 as Britain's first colony, was about to become self-governing. Its leaders saw that the telegraph venture could open up the island and give it an important position in the wider world. Within hours they confirmed an offer guaranteeing the interest on £50,000 worth of bonds, an immediate grant of fifty square miles of land, with more to follow once the transatlantic telegraph was in place, and a donation of £5,000 towards the building of a bridle path across the island on the line of the land telegraph.

The necessary charter was not then confirmed, but Cyrus Field was hopeful enough to leave St John's after three days, returning to New York to buy a steamer, the *Victoria*, for the new company. White and Dudley Field were left behind to deal with the legal arrangements. The old Electric Telegraph

Co. had to be dissolved, and a new one established which would bear a longer title – the New York, Newfoundland & London Electric Telegraph Co. – to take account of its wider ambition. Dudley Field had drafted the new company's charter while on the voyage to St John's. It stated categorically that a telegraph line would be established between America and Europe via Newfoundland. No other company was to be allowed to land cables in Newfoundland or its dependencies, which included Labrador, for fifty years.

The new company, seemingly through the efforts of Gisborne, also acquired generous concessions on Prince Edward Island and a twenty-five-year exclusive right to land cables on the coast of Maine, giving it in total concessions covering 2,000 miles of coastline. These valuable rights gave considerable power to the company in the following years, when other parties were attempting to move into the world of long-distance deep-sea cables.

After some weeks of negotiation, the agreement was finally approved by Newfoundland's legislature. Dudley Field and Chandler White threw a banquet in St John's for forty invited guests to mark the signing of the deal. The atmosphere was one of excitement and optimism, as the isolated colony looked forward to being put on the map. Sensing that they had long been neglected, Newfoundlanders eagerly anticipated the increased trade and notice that the cable would bring. Bryan Robinson, a lawyer and member of the executive council, addressed the diners on behalf of the island authorities: 'The establishment of an electric telegraph from St John's to the United States, to be connected with England by powerful steamships, involves consequences so stupendous to the interests of this country that it is not easy to know where to begin in the enumeration of them.' He looked forward to the finest steamers in the Atlantic regularly visiting St John's. Robinson praised the company, a group of 'enterprising and wealthy American citizens', but did not overlook the efforts of the engineer, Frederic Gisborne. 'Let it not be forgotten that if it had not been for the scientific acumen of Mr Gisborne, who devised the scheme, and with indomitable perseverance in spite of checks, reserves and disheartening disappointments, still adhered to it, and induced the present corporation to embark in the project.'

Gisborne described the occasion as 'a very proud moment in my somewhat eventful career'. He looked forward to something more than the present scheme promised, to a time 'when we can whisper mutual words of friendship through the mystic voice of electricity, even though oceans roll between us'. The Atlantic cable had already gripped public imagination around the globe, and there was rising expectation that the whole world would soon be in communication. Chandler White, though, the retired businessman and vice-president of the company, while agreeing with Gisborne to a point, had quite a different approach. White told his guests that the cable should be seen 'in the light of a great commercial enterprise, directed to the realization of profit, the acquisition of gain, the accumulation of wealth'. He knew that the undertaking was a special one in many ways, but believed firmly that it was driven above all by 'the mercantile principle'. This was the first inkling of a friction, not just between two men, but between the contrasting approach of engineer and businessman, of the tension between idealism and pragmatic commerce, which would both bedevil and propel the Atlantic project for a decade to come.

The new company brought an immediate benefit to the colony, for it re-opened the old Electric Telegraph Co. office and used $50,000, which Cyrus Field had sent from New York, to settle the wages and debts owed from Gisborne's work the previous year. White and Dudley Field then returned to New York, arriving there at the beginning of May, when the first formal meeting of the company took place. In a few minutes, $1.5 million dollars was subscribed. Morse was nominated as advising electrician, and was given a tenth interest in the company for $10,000. As well as his technical experience, Morse was able to offer some sway over the overland telegraph companies on the North American mainland, whose co-operation was vital to the new company's success. Maury declined a share in the business, as it would have prevented him from using his influence in official circles when he already had 'the success of the enterprise much at heart, and would be glad to do what I rightfully may to forward it'.

All this had been achieved within four months of Field's first meeting with Gisborne. But as his brother, Revd Henry Martyn Field, later wrote: 'Well was it for them that the veil was not lifted, which shut from their eyes the long delay, the immense toil, and the heavy burdens of many wearisome years.' Cyrus Field imagined that he was risking a few thousand which he could easily afford, on a venture which may come to nothing. He could not have thought that over the following twelve years he would spend most of his time re-crossing the Atlantic in a sometimes desperate attempt to keep the telegraph project alive, and ultimately gamble everything he owned.

Even before the transatlantic cable was commissioned, there was much to organise on the American side of the ocean. It was no small undertaking to complete the land line across Newfoundland. The forty miles of telegraph line so far constructed were in the inhabited lowlands. The most difficult stretches remained to be done. There was also the Cabot Strait to cross. The government of Nova Scotia, fearing that they would be left out in the cold, had relented on the landing rights and the chosen route was now Cape Ray to Cape Breton Island, a distance of sixty nautical miles at up to 300 fathoms. A 140-mile link across Cape Breton would also be needed.

Chandler White, who weeks earlier had been enjoying the views of New York from his retirement retreat, found himself appointed general manager of operations in St John's. Matthew Field was placed in charge of construction of the Newfoundland land line, while Gisborne was designated consulting engineer. Within a month of the triumphant inaugural dinner, Gisborne had resigned from the company, for reasons not fully explained. Most likely he resented losing his autonomy and having to answer to Matthew Field, a man with little knowledge of telegraphs. Gisborne, though still only thirty years old, had a wealth of experience in this new industry.

Matthew Field recruited 600 labourers for the Newfoundland job. The land telegraph was to be 400 miles in length, following the coast as it was near impossible to provision such a body of men on any inland route. Supplying building materials and food was difficult enough by sea. The great camp of huts and tents was serviced by the *Victoria*, delivering 'barrels of pork and potatoes, kegs of powder, pickaxes and spades and shovels, and all the implements of labor', all of which had to be man-handled inland to the camp. Henry Field saw the romance of this expedition, for it was 'a wild and picturesque sight to come upon their camp in the woods, to see their fires blazing at night while hundreds of stalwart sleepers lay stretched on the ground. When encamped on the hills, they could be seen afar off at sea. It made a pretty picture.' It was expected to take a few months, or at the worst be finished by the following year. In fact it was two and a half years before it was completed. As Cyrus Field later wrote, 'It was a very pretty plan on paper. There was New York and there was St John's, only about 1,200 miles apart. It was easy to draw a line from one point to the other, making no account of the forests and mountains and swamps and rivers and gulfs that lay in our way. Not one of us had seen the country or had any idea of the obstacles to be overcome.' At one point during 1855 he ventured to ask of his brother how many more months the line would take. Matthew replied:

How many months? Let's say how many years! Recently, in building half a mile of road we had to bridge three ravines. Why didn't we go round the ravines? Because Mr Gisborne had explored twenty miles in both directions and found more ravines. That's why! You have no idea of the problems we face. We hope to finish the land line in '55, but I wouldn't bet on it before '56, if I were you.

The telegraph once completed ran from a cable station at Point-au-Basque, on the western extremity of Newfoundland, to Trinity Bay, facing the Old World on the east of the island.

The Atlantic project had been an almost exclusively American enterprise, financed and driven by a small group of men. 'Our little company raised and expended over a million and a quarter of dollars before an Englishman paid a single pound sterling,' recalled Cyrus Field. Any one of the projectors was able to write a large cheque as further funds were needed, so there was no necessity even to ask the American public to invest. By the end of 1854, half a million dollars had been spent, some of it with Peter Cooper's wire-drawing business, which supplied the overland sections of the telegraph.

In one respect, though, the Americans could not be self-sufficient. England was the only possible source of submarine cables. British dominance in this area was partly a result of their monopoly on the insulating material, gutta percha. More importantly, much of the expertise accumulated during the short life of undersea cables was concentrated in Britain.

Cyrus Field therefore set out again to London, to inspect samples of cable made for him by Kuper & Co. using copper cores from the Gutta Percha Co. His only other contact there was John Watkins Brett, who had become head of the Magnetic Telegraph Co. Gisborne, by then on travels around the United States, had furnished Field with a letter of introduction to Brett.

Brett accompanied Field to the factories in London where cable was made, and offered advice on the best structure for a cable. He also invested in the new company and continued to represent its interests in Britain once Field had returned to the United States. While Field was in London on this first telegraph trip he took the opportunity to meet Isambard Kingdom Brunel, Charles Bright and Wildman Whitehouse, and to seek their advice on the transatlantic scheme. Brunel had no direct involvement with submarine telegraphs before or after Field's visit, though he offered some thoughts on the subject to the American. He is also supposed to have taken Field to Millwall, where his gigantic ship *Great Eastern* was under construction, pointing out: 'There is the ship to lay the Atlantic cable!' This story has the ring of a myth about it. Henry Field, from whom the account originated, says in another context that once the transatlantic cable was a success, 'a host' had sprung up 'to claim the honor'. Although Brunel was long dead by the time his great ship was converted to cable-laying, there were those who wished to have his name associated more closely with deep-sea cables. But it was Bright and Whitehouse who were of more immediate use to Field, and both later held engineering positions with the Atlantic Telegraph Co. Although Whitehouse's limitations were soon to be painfully exposed, both he and Bright in 1855 were 'full of the ardor of science' and zeal inspired by 'the prospect of so great a triumph'.

The submarine cable from Newfoundland to Cape Breton was to be laid from Port aux Basques by the barque *Sarah L. Bryant*, supervised by a British engineer, Samuel Canning. Field knew that there had been a problem in commissioning machinery to pay out the cable from the ship for this expedition, as Kuper's were unable to supply it. Making the cable core, insulating it and then laying the cable were at that time three separate operations, carried out by different parties. Field must have believed that construction of the paying-out machinery was in hand, for he went ahead in chartering a luxury coastal steamer from the Charleston Line. This vessel, the *James Adger*, was needed to tow the *Bryant* across the Cabot Strait. The *Adger*, expensive at $750 a day because she was equipped with the most modern facilities, was essentially chosen for a social purpose. Fifty spectators, families of directors and other society figures, were invited along, embarking from New York early in August 1855.

Field had made a huge miscalculation. He and the other directors of the revived Newfoundland company had spent such a time reassuring one another that the scheme must succeed, that they had overlooked the great difficulties still before them. The presence of fifty trippers in search of a carnival added to the problems, and to the embarrassment when the expedition failed. The weather turned rough and the *Bryant* could not be found. The *Adger* went instead to St John's, where a round of dining and dancing took place. The two ships finally met up in Port aux Basques, where the *Bryant* was found to have been damaged by storms and in need of repair. Although she carried the cable-laying machinery, it had not been assembled. Other work remained to be done. No site had been identified where the cable could be brought ashore, so that a cable station had to be built from scratch in Cape Ray cove. The company's steamship *Victoria* was on hand to help but had difficulty in landing the materials needed. There was further delay when fog descended for two days. When finally the *Adger* took the *Bryant* in tow, there was a misunderstanding, the *Bryant* lost her anchor and the vessels collided. The *Bryant* came off worse, narrowly avoiding sinking, while the *Adger*, her captain in a rage at having to take orders from Field, steamed away from the scene.

After the *Bryant* had been repaired, Captain Turner of the *Adger* was involved in a furious row with Peter Cooper, on his first ever ocean voyage. Turner had been instructed by the engineer, Canning, to steer away from the port while keeping a flag on the cable station in line with a white rock on the mountain behind. In this way he would maintain a correct course. It became apparent to Cooper that Turner was not following this line, but the captain would only insist 'I steer by my compass'. Cooper had a lawyer on board write to the captain, warning him that the company would hold him responsible for the loss of the cable if he did not change his route. 'He then turned his course and went as far out of the way in the other direction.'

When the wind turned into a gale, twenty-four miles of the cable had been laid, although they were only nine miles out from shore. The ship's irregular course meant that there were great kinks in the line. These were likely to affect the future working and maintenance of the cable, assuming even that sufficient wire remained to finish the job. The total distance to be covered was sixty nautical miles, and although seventy-four miles of cable were laid, they were nowhere near their destination. Worse, Captain Turner refused to moderate the speed of his ship despite pleas by the cable-laying vessel. In the end, to save herself from being dragged under, the *Bryant* was forced to cut the cable, which was lost. Peter Cooper later placed the blame firmly with the stubborn captain. 'We had spent so much money, and lost so much time, that it was very vexatious to have our enterprise defeated by the stupidity and obstinacy of one man.' Turner, 'one of the rebels that fired the first guns on Fort Sumner', was subsequently killed in action fighting for the Confederacy.

Canning and Cyrus Field were distraught at the expedition's failure. All that could be done was to offload the unused length of cable in Sydney, Nova Scotia, and take the revellers back to New York. The company had spent $350,000 and had nothing to show for it. Peter Cooper tried to cheer Field, as he was optimistic that the lessons learned had been valuable ones. Success could not be far away. They now knew that it was impossible to use a vessel such as the *Bryant*, without its own steam power, on such a mission. The cable-laying ship needed its own steam, to better regulate its motion and speed, manoeuvring as the sea rose and fell so that the strain on the telegraph line was never too great nor too small.

A year after the very public defeat of 1855, the links from St John's were finally completed, with little fanfare. The ship *Propontis* had been fitted out for the purpose. Samuel Canning took charge, assisted by a young electrician called Charles de Sauty, later to be superintendent of the Newfoundland telegraph station during its short life of 1858. Again, seventy-four nautical miles of cable were laid, but this time successfully joining Cape Ray to Cape North on Cape Breton Island. Canning also put down a new twelve-mile stretch from Prince Edward Island to New Brunswick. The Nova Scotia telegraph company made the link from there to Cape Breton, only one and a half nautical miles in length. These small submarine connections brought the North American continent into direct communication with Newfoundland for the first time, via the 400-mile road and telegraph crossing the island, and a further 140-mile line across Cape Breton Island. The British mainland was already linked with Ireland and with continental Europe, so that there was only one further step needed to connect most of the world into one telegraph system. That one remaining gap was the Atlantic Ocean.

2

The Great Feat
of the Century

The new telegraph between St John's and New York had swallowed a million dollars, but it ran smoothly and without interruption. The Atlantic projectors had learned a lot by 1855 and their work could be counted a success, as far as it went. The next stage, though, was daunting. It would be an operation on an unprecedented scale and at vast cost, laying almost 2,000 statute miles of cable at depths of two miles or more. No one had yet been able to answer Lieutenant Maury – could there be 'a time calm enough, the sea smooth enough, a wire long enough, a ship big enough'? The *Adger*, after all, had failed on her much smaller mission. But the ambitious venture excited the public and challenged the engineers, and did not yet unduly bother the financiers.

Ten years earlier, the idea of an Atlantic cable had seemed to Morse as wildly improbable as air travel. But a decade before that, land telegraphs had still been in the realm of theory, and it was not until 1845, some time after Morse had sent signals across New York harbour, that Cooke and Wheatstone's single needle apparatus made overland telegraphs cheap, reliable and widespread. By 1850 all the main towns and cities of Britain were linked into the telegraph network.

The electric telegraph had not been conceived by Cooke and Wheatstone. In fact, the idea of using electricity as a means of signalling was around before 1750, though the theory could not then be translated into much that was practical. The first functioning telegraph was set up by Francis Ronalds, a distinguished amateur scientist, in his Hammersmith garden in 1816. Ronalds laid eight miles of wire underground, insulated in glass tubes, and using a static charge. He was able to transmit signals, but too slowly, and the system was not sufficiently robust or reliable for commercial operation. Nor did Ronalds convince anyone that the device might actually have a useful purpose. The Admiralty politely rebuffed his suggestion that the telegraph may assist their operations. But later electricians acknowledged the part that Ronalds had played in laying the ground for telegraphy. After 1831, Michael Faraday's discovery of electromagnetic induction made further advances possible.

Charles Wheatstone was professor of Experimental Philosophy at King's College, London, when he met William Fothergill Cooke in 1837. Wheatstone, fascinated from an early age by sound, had started his career as a musical instrument maker. Later he turned to experiments on light and optics. Appointed to the chair at King's in 1834, he started work on the transmission of electricity along wires, and was the first to measure the speed of electricity – inaccurately, as it turned out, at 250,000 rather than 186,000 miles a second

– by experiment in the college vaults. He was also quick to recognise the importance of Ohm's law as a foundation of electrical engineering. This was the simple rule, first stated in 1827, about the relationship between electromotive force, the resistance of conductors, and current.

Cooke had been invalided out of the Indian Army and was considering a career in medicine when by chance he witnessed telegraph experiments in Germany. He had the idea of developing the electric telegraph in tandem with railways, which were also in their infancy. The scheme had great attractions – telegraphs could run conveniently alongside railway lines, and were exceptionally useful to railway companies in organising their operations. Cooke threw his energies into understanding telegraphy and developing new instruments. When he met Wheatstone, they pooled their experiences and took out their first joint patent in 1837. In 1838 a telegraph was at work alongside the Great Western Railway between Paddington and West Drayton, in Middlesex. Wheatstone, despite his theoretical interests, was particularly skilled in developing practical applications. Cooke, like Morse, did not come from a scientific background, but was rather a visionary who spotted early the tremendous potential of telegraphs. The partners worked together for a time to develop instruments, with Cooke handling the business matters for which he had great talent. Wheatstone's first instrument had five needles, pointing to letters of the alphabet around a dial, needing six wires to make it work. A breakthrough came when it was realised that trained operators using a code and a single wire could send and receive messages much more speedily. Using a single wire also made the system less prone to break down. The code generally adopted was Morse's, developed in 1838.

Cooke and Wheatstone fell out spectacularly, just at the time when their electric telegraph was sweeping across Britain, over who deserved most credit. After a quiet start to life, the telegraph had suddenly leapt into the public's consciousness. The Berkhamsted town surveyor, John Tawell, had been arrested in London after fleeing a murder scene in Slough in 1845. A description had been telegraphed to Paddington, and Tawell was quickly caught and later hanged for his crime.

As the cable spread, it became clear that it was far more than a means of catching fugitives from justice or to run railways. In the United States, a telegraph from Washington to Baltimore in 1844 marked the start of a rapidly developing network. Cables were also appearing across the continent of Europe. It was in central Europe that a remarkable career was built on the sale of information. Julius Reuter, after several false starts with his news agency, took advantage of a public telegraph between Berlin and Aachen which opened in 1849, to deliver news and prices between Berlin, Vienna and Paris. Where there were gaps between the new telegraph systems of France and Belgium, he rushed the messages by pigeon, which was faster than trains. Political and other news was an important commodity, but Reuter's main business was trade information, prices and rates, where speed and accuracy were of the essence, and which he sold to newspapers and to commerce. He made a huge success of being ahead of the game, and moved his base to London in 1851, on the eve of the opening of the Channel cable.

The Brett brothers saw some other possibilities for international cables. When trying to raise support for their transatlantic telegraph in 1845, they pointed out to the Prime Minister, Sir Robert Peel, the advantages of instantaneous communications with all parts of the Empire. If the government did not then see how the cable could help rule the world, it did so soon afterwards when inland telegraphs proved invaluable in deploying troops and police against the Chartist threat.

The Bretts found fame as promoters of the first international submarine cable, laid between Dover and Cap Gris Nez near Calais, late in August 1850. They had obtained a ten-year concession, or monopoly, on the route. The Gutta Percha Co. supplied twenty-five nautical miles of cable. As the line was short and weighed only five tons, it could be wound around a drum and laid direct from that, using weights. This was done in a day, and messages successfully exchanged between the shores. The following day, though, there was no sign of life in the telegraph, and a French fisherman was blamed for pulling up the cable with his anchor and cutting it to take as a souvenir.

While Julius Reuter, by his pre-emptive move to London, showed faith that the cable would be completed the following year, many others did not. Ominously for the Atlantic projectors, it became clear that wild public enthusiasm could quickly give way to despair and suspicion. But compared to the Atlantic scheme, the Channel project was small fry. It was saved in 1851 by one man, a railway engineer, Thomas Russell Crampton, who raised or invested most of the £30,000 needed and also improved the design of the cable. He set a pattern for later telegraphs in using an outer protective layer of galvanised iron. The cable was successfully laid in the autumn of 1851, and remained in constant operation after that. Confidence was reborn, and, after some setbacks, other submarine lines quickly followed. By the end of 1853 there was a telegraph from Portpatrick in Wigtownshire to Donaghadee in County Down, giving the

British government a valuable tool for its rule of Ireland. There were also new connections to the continent – lines from Dover to Ostend, and from Orford Ness in Suffolk to the Netherlands. The longest of these was 100 miles, the deepest 160 fathoms.

The success of these early submarine cables disguised some of their inherent flaws. On a short or shallow line, faults did not necessarily stop the telegraph from working. In the depths of the Atlantic, though, they would be fatal. At the root was the quality of the cable. Wire-drawers, even within the same factory, could not make wire to a consistent gauge or standard. But cable engineers paid little attention to the wire's purity and electrical performance, for it was assumed that all copper wires behaved the same.

Nor was it easy to apply the gutta percha insulator evenly around the conductor. Gutta percha, an 'imperishable sub-aqueous insulating material', was heralded as the nineteenth-century wonder material. It was not a new discovery, but found a use only after Faraday during the 1840s recognised its excellent insulating properties. Electricians soon found that it worked much better than India rubber or any other compound on marine cables. Derived from the latex of gutta trees found only in the Malay peninsula, gutta percha is a natural plastic which can be shaped when hot and stays flexible as it cools. Once applied as an insulator to submarine cables, it needed to be stored in water to retain its remarkable properties. Its supremacy lasted for a century, until the advent of polyethylene-based synthetics, handing Britain, which had a monopoly on gutta percha, a long-lasting stranglehold on the production of undersea cables.

While the material itself was seen as a godsend, there was still some way to go in improving the way it was applied to the cable. With land telegraphs, most of which were carried on overhead lines, insulation was hardly an issue. The cable was insulated by glass or earthenware holders at points where it met a telegraph pole. For a submarine line, though, the insulation had to be of consistently high quality to prevent any contact between the copper conductor and the water all around.

The insulation problem turned out to be much more than a matter of waterproofing. Even when the quality of the cable was excellent, and the insulation perfectly sound, undersea cables simply did not function as expected. Messages would not pass down the line at anything like an acceptable speed without breaking down into a chaotic jumble. This electrical phenomenon came to be called 'retardation of the signals', sometimes known as 'induction'. Retardation was at the root of problems with long cables, and until it could be understood and its effects overcome, the Atlantic cable could never work.

The contrast with land telegraphs was striking. By the early 1850s, overland telegraphy had achieved a measure of sophistication. Signals could be transmitted more or less automatically and at relatively high speeds. Experienced clerks understood the sound of the incoming signal almost as a language and could read off a message just by listening. Once printing receivers were introduced, the clerk no longer had to be there constantly to note down the message. Little attention needed to be given to what was happening in the conductors themselves or in the insulating envelopes. If it proved difficult to transmit intelligible signals, electrical relays could easily be inserted to boost them. This meant there was no limit to how far a telegraph line could be extended. Morse had suggested, as early as 1837, the use of relays:

Suppose that in experimenting on twenty miles of wire we should find that the power of magnetism is so feeble that it will but move a lever with certainty a hair's breadth; that would be insufficient, it may be, to write or print, yet it would be sufficient to close and break another or a second circuit twenty miles further, and this second circuit could be made in the same manner to break and close a third circuit twenty miles further on, and so round the globe.

For most electricians, any problem of signalling came down to 'strength of the electricity', in other words, current. It was believed that if enough current were transmitted through the wires, there could be no difficulty in operating a receiver.

Retardation showed itself very quickly on the first submarine lines. The difficulty lay not in the speed of electrical currents in the cables, which was known to be very rapid indeed, but with the rate at which letters and words could be transmitted. During laying, the first Channel cable in 1850 was tested only for electrical continuity. When it was complete, and the time came for signals to be received from Dover, it was assumed that something was amiss with the operator:

Letters came, but they were so mixed that it was in many cases impossible to make any sense out of them ... The more the operator tried to control the letters the more erratic they became. At last it was suggested that the success attending the laying of the wire had caused the champagne to circulate so freely that the persons in the shore station at Dover did not know what they were doing.

It was not alcohol, but the retardation phenomenon, which was causing the problem. It was quickly confirmed that messages were being correctly sent, yet only unintelligible, chaotic signals could be received. The immediate solution was to restrict the rate at which operators worked, but this limited traffic so much that it threatened the line's commercial viability. The electrician Willoughby Smith lamented many years later that the phenomenon of electrical induction had not been understood sooner. Looking back at the earliest undersea cables, Smith could see that the blame for their poor performance fell upon failures in scientific understanding as well as inadequate testing and quality control.

Faraday had discovered a general phenomenon called 'the capacitance effect' in 1838, and was able to explain how this worked to hold up underwater signals on a telegraph line. The problem had become evident again on the Channel line of 1851, and then in the cables, 100 miles in length, connecting Orford Ness with Holland in 1853. Faraday realised that a submarine cable, made from a central copper conductor surrounded by an insulating envelope of gutta percha, armoured on the outside with iron wire rope, formed an electrical capacitor. A capacitor is a device that can store electrical charge, the first practical example of which was the Leyden jar. So, when a pulse of electricity was sent into a telegraph cable, there were two processes at work: an electrical current through the core, and an accumulation of charge in the capacitor. The net effect was that it took some time before the results of a pulse applied to the input end of the cable became evident at the output. If successive pulses were too close together, confusion would result. It soon became plain that the problem of retardation was much more profound than previously understood, and that stronger currents would not solve it.

Edward Bright measured the effect in the early 1850s, clocking the speed of subterranean and submarine currents at less than 1,000 miles per second on a length of cable which was about 500 miles long. As the problem evidently intensified as the cable lengthened, electricians began to fear that it might prove to be physically impossible to send a message across the distance of the Atlantic. William Thomson wrote in 1854 that, while he had no doubt about the feasibility of successfully making and laying an Atlantic cable, a rapid rate of signalling may never be achieved. On the eve of the first transatlantic cable in 1857, scientists could attempt to describe the phenomenon but were no closer to a solution:

> When the wires are enclosed in a compact sheath of insulating substance, such as gutta percha, and are placed in a moist medium, or in a metallic envelope, the influence of induction comes into play as a retarding power to a very sensible extent. So soon as the insulated central wire is electrically excited, that electrical excitement operates upon the near-at-hand layer of moisture or of metal, and calls up in it an electrical force of an opposite kind. These two different kinds of electricity then pull at each other … through the intervening layer of impenetrable insulator, and each disguises an equivalent portion of the other, that is, holds it fast locked in its own attraction and so renders it valueless as an agent of extraneous power. The inner electricity keeps the outer induced force stationary upon the external surface of the insulating sheath. The outer induced electricity keeps a certain portion of the inner, excited

influence, on the interior surface of the insulating sheath as a charge, and so prevents it from moving freely onward on its journey as it otherwise would. The submarine cable is virtually a lengthened out Leyden jar ... a bottle for the electricity, rather than a simple channel or pipe open freely at both ends.

While doubts persisted, submarine cables received a great boost from a national emergency, the Crimean War, which supplied the impetus and the funding for an ambitious scheme. The Crimea has been called the first modern war, and it was the telegraph that made it so. The Royal Engineers set up a circuit of twenty-one miles of cable and eight telegraph offices on the Crimean peninsula, but this was of limited use without a submarine line to link to the outside world. The cable was an instrument of war, a means of managing the conflict, but it was also to be the medium through which news of carnage and incompetence found its way back to the public of France and Britain.

The Crimean cable was in two sections, laid by Robert Stirling Newall, who had made his name as main contractor on the 1851 Channel cable and become the most prominent cable-maker and layer of the time. The main cable connected Varna in Bulgaria with Balaklava on the Crimea, a distance of 270 nautical miles and to a depth of 950 fathoms. A shorter line brought Balaklava into communication with Eupatoria, north of Sebastopol. Both these lines, commissioned by the British government and rapidly installed without armouring in 1855, were considered temporary, although they performed well until the war ended the following year. Newall also laid a line for the Ottoman government from Varna to Kilia, in the mouth of the Danube in south-west Ukraine, 150 nautical miles long and up to 500 fathoms deep, which soon failed.

Up to this point, there had only been one successful cable laid in a depth of more than 150 fathoms (about 300m). This was between La Spezia in northern Italy and Corsica, a distance of seventy nautical miles and to a depth of 325 fathoms, completed in 1854 by the Brett brothers and others for the Mediterranean Telegraph Co. This line worked without interruption for ten years. In 1855, the Bretts tried to complete the 130-mile link from Europe to North Africa with a cable from Sardinia to Algeria. The depth turned out to be more than double the estimate of 800 fathoms. After two failures, the expedition ran short of cable a few miles from Galite Island, off the Tunisian coast, and the campaign was abandoned. Newall and his partner Charles Liddell, their reputations enhanced by their work in the Crimea, were employed by the Mediterranean Telegraph Co. in 1857 for another attempt on the line in the autumn of that year, and this time were successful.

So, by late 1856, as Field and his associates prepared their assault upon the Atlantic, any lessons to be gleaned from their own and others' experiences of deep-sea cables were far from clear. In total thirty-eight submarine telegraphs had been laid, and many of those were working well. But the ones which performed best were short and shallow. Although the Mediterranean cables were beginning to experience some success, the Italy-Corsica line was still the only one working at more than a quarter of a mile in depth. Furthermore, the retardation question remained unresolved.

It was not in Field's nature to give up, and a great deal had already been committed to the Atlantic scheme. The Atlantic directors had to rely on assurances from their technical advisors, the electricians, who were naturally also inclined towards optimism. This was the first generation of telegraph engineers, a small group of mainly very young men who laid the foundations of what would become known as electrical engineering. Some had come out of the telegraph industry, others had academic backgrounds. As with any new technology, experience was necessarily very limited, and they struggled to understand the electrical questions which long-distance cables threw up. There was only so much that could be achieved by experiment in the factory or laboratory, so the young electricians relished any chance to carry out research on the biggest stage of all, out at sea.

For a time, though, electrical problems took a back seat. The immediate issues which needed resolving, and the ones which most fascinated the general public, were mechanical ones. It was easy to understand the challenge of handling and laying the monster cable across a vast ocean. The plan in 1857 was to start putting down the telegraph from mid-Atlantic, where the two sections would be spliced and the ships steam in opposite directions. This meant that the line could not be too heavy, for when laying started there would be a length of up to five or six miles of cable between the ships, a huge mass not immediately supported by the ocean bottom. This vast floating weight endangered the cable, and perhaps the ships themselves. Then again, the cable must not be too insubstantial, for it had to sink to the bottom under its own weight. Friction, as well as the line's own buoyancy, would act as a brake, up to a point. A line that was too light would reach the bottom, but it would not be laid straight, as currents would move it from its direct route and introduce damaging kinks.

Once the ideal weight had been established, neither too heavy nor too light, there was a question of how to deal with the cable's bulk. How was 2,500 miles of armoured telegraph cable to be stowed and handled? It was suggested as early as 1857 that 'Mr Scott Russell's leviathan ship, the *Great Eastern*' would be ideal for the job. Brunel's mighty ship, an eighth of a mile in length, was under construction in John Scott Russell's yard at Millwall on the Isle of Dogs. Unfortunately, though, the *Great Eastern* was incomplete – she was floated only in 1858, and finished in 1859 – so untested. 'It must have proved its own armour, before it can be discreet to ask it to undertake a campaign for an ally.' Two gigantic and costly experiments, the cable and the leviathan, could not be risked together.

Without the *Great Eastern*, the only possible solution was to use two ships. But even the largest vessels in existence could not carry 1,250 miles of cable if it weighed more than a ton a mile. It was essential that the cable was as strong as possible, to guard against accident – but while it could not be too strong, it may be too bulky, too heavy or too light. It must also be flexible enough to coil in the factory and in storage, and to be easily handled through the sheaves of the paying-out machinery. The copper core had to be of sufficient quality and purity to transmit long-distance signals, and the insulation needed to withstand the weight and forces of the Atlantic deeps. The specification finally agreed was for a pure copper wire, twisted from seven strands and about a sixteenth of an inch in diameter. It was designed to stretch by twenty per cent without breaking, and even if strands were broken at different points could continue to conduct electricity. When coated with three layers of gutta percha, the diameter became about three-eighths of an inch, rather larger than previous submarine cables.

By the time the Atlantic crossing was planned, submarine cable manufacture had become much better organised and regulated, in the hope of avoiding some of the calamities of the early projects. Once coated with gutta percha, two-mile lengths of the core were immersed in water and carefully tested for faults. These tests were made with a galvanometer, a sensitive instrument which could detect whether a current passed through the wire. After being joined into sections of about 100 miles, the cores were sent to another factory for armouring. A layer of hemp and tar protected the gutta percha from damage by the outer coating, the armour itself.

The armour's main function was to protect the cable from 'mechanical violence', although it also gave scope to make the line heavier or lighter, as appropriate for particular circumstances. It shielded against damage during paying out, and guarded the cable from accidental harm by anchors or fishing lines or rocks once it was laid. On each submarine telegraph there were two, more heavily armoured, shore-end cables which were spliced to the main cable some miles out at sea where the risks of accidental damage became much lower. For the Atlantic cable, there were to be ten miles of shore-end cable at the Trinity Bay end, and fifteen miles off the coast of Ireland. These parts of the cable weighed seven tons a mile, more than seven times as much as the main section. The engineers expected that the iron coating of the main cable would eventually be consumed by rust, but once the cable was on the ocean floor this did not matter, as the gutta percha-coated core was designed to withstand the pressure of water and to survive independently.

Charles Tilson Bright.
(*Cable & Wireless Archives, Porthcurno*)

Charles Bright and Wildman Whitehouse, who had made Field's acquaintance in London in 1855, were appointed to the Atlantic company: Bright as engineer and Whitehouse as electrician. Bright, only twenty-five years old in 1857, had started his telegraph career ten years earlier under Cooke. He rose to become engineer of the Magnetic Telegraph Co. by the age of twenty, and was involved in laying the first successful cable between Scotland and Ireland the following year. He made a large personal investment in the Atlantic scheme at the outset. Wildman Whitehouse, a Brighton surgeon long fascinated by electricity, was associated with John Brett during the laying of the Channel cable. By 1854 he had virtually abandoned medicine to devote his time to electrical experimentation.

Even before they joined the Atlantic company, Bright and Whitehouse had been preoccupied with the twin problems of whether it was indeed possible to transmit an electrical current across 2,000 miles of ocean, and if so, whether signals could be sent and received at a commercially viable rate. They spent much of the middle years of the 1850s working on these questions.

Experimenters would seize on any chance for trials. In 1855, when two long cables had been completed in the Greenwich factory, a line of 1,146 miles could be constructed by linking them together. The results were exciting. Until then, there had been genuine doubts that an electrical signal could be transmitted at all along a cable of such length. Within moments there came an answer: a trace appeared at the far end of the line. It could no longer be denied that the signal would travel 1,000 miles at least. But the signs were not entirely good. The current, transmitted along the cable as a series of beats, was received as one long continuous line.

In the autumn of 1856, Bright, Whitehouse and Morse set out to prove without doubt that it was possible to send a signal down 2,000 miles of telegraph line, and do it at a speed which would make the Atlantic cable economically viable. Morse had emphasised all along that the Atlantic scheme must be practicable, that is, that it could be achieved technically, but he believed it should also be practical; in other words, sensible and to the advantage of those proposing it. The American travelled to London to meet Bright and Whitehouse, whom he described as 'that clear-sighted investigator of electrical phenomena', at the offices of the English & Irish Magnetic Telegraph Co. Bright had arranged a night-time experiment in which ten gutta percha-insulated underground lines, each of more than 200 miles, would be linked. This trial would mimic the transatlantic cable as closely as possible, not only in length, but also because subterranean lines closely resembled submarine cables in their electrical properties.

These underground cables were already in commercial use, so that Morse and his 'active and agreeable collaborators' had to conduct their trial at dead of night, when the office was closed. Bleary-eyed after a full night of testing, during which the electricians were able to transmit up to 270 signals a minute in conditions which were far from ideal, Morse concluded that the distance was in both respects – of practicability and practicality – viable. Having slept on it, he wrote to Field, estimating that eight to ten words a minute, or twenty messages an hour, could be sent between Ireland and Newfoundland. This amounted to 480 cables a day, and that on a single wire. Morse thought that this rate could be doubled by further improvements in signalling, in particular a more refined code, but, he concluded, 'the doubts are resolved, the difficulties overcome, success is within our reach and the great feat of the century must shortly be accomplished.'

Doubts persisted, though, about managing this rate of transmission on the line itself. This was a real threat to the cable's profitability. The slower the speed at which messages could be sent, the more astronomical would be the cost of each telegram, and the lower the return to shareholders. After Morse's triumphant note to Field, Bright and Whitehouse continued their experiments. The best explanation for delayed signals that they could advance was that the conducting wire did not discharge itself between beats. 'The current moved so sluggishly and unwillingly in this protracted and induction-encumbered channel, that one transmission was not able to clear itself off, before the next was pressing upon its heels, and in confusion with it.'

A solution to the retardation dilemma began to emerge, not from the practical experiments of Whitehouse, but from an altogether more theoretical realm. Key to these developments was William Thomson, later Lord Kelvin, who built on Faraday's work. Thomson was professor of Natural Philosophy at the University of Glasgow, and an associate of R.S. Newall and his partners, Lewis Gordon and Charles Liddell. Thomson convinced Gordon that 'scientific deductions from established principles' were a reliable and economical method of advancing submarine telegraphy – much more so, in fact, than expensive practical trials from which even the best engineers could draw mistaken inferences. Working purely from theory, Thomson produced graphs showing that a pulse that started out more or less rectangular in form would emerge from an insulated cable rounded and elongated.

Thomson's theoretical work proved that the speed of signals depended on three things – the length of the cable, and the properties of the copper conductor and of the insulator. Of these three variables, cable length was the most significant. Retardation of the signals was, he concluded, in proportion to the square of the cable's length – Thomson's 'Law of the Squares'. It was possible to increase the speed significantly by making the copper core larger, but this brought with it other difficulties. It made the cable bulkier and heavier, so that bigger ships, heavier paying-out machinery and larger cable-making machinery were needed. All these had an impact on costs, and brought further questions about the economic viability of transoceanic telegraphy. Otherwise, physically, there was little else to do to improve the cable. Reducing impurities in the copper core would increase conductivity, but the gains were small. As for the insulator, nothing had been found which could come close to the performance of gutta percha.

While Thomson and others in his circle were convinced by the 'Law of the Squares', Wildman Whitehouse would not accept it. He went so far as to produce experimental results in support of his view, at a meeting in 1856 of the prestigious British Association for the Advancement of Science. Thomson could not verify his law by experiment because, while input signals could be timed precisely, it was much more difficult to measure exactly the exit signals. Later it emerged that Whitehouse's measurements were flawed. It was not until 1859 that Fleeming Jenkin, who worked closely with Thomson, produced a clear proof of the 'Law of the Squares'. By that time, Thomson and Jenkin, both experienced cable engineers as well as able theoreticians, had realised the enormous value of merging their abstract and practical knowledge, science with technology. They saw ways of designing instruments which would vastly improve signalling speeds. Thomson showed, through theory, that by accurately timing a mixture of positive and negative pulses, a sharp, readable message would result. Jenkin verified this, confirming that several words a minute could be sent through cables thousands of miles in length.

But all this came too late for the 1857 cable.

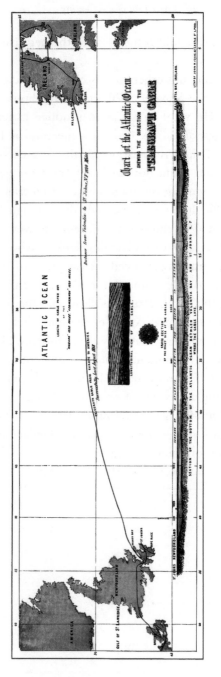

1 Route of the 1858 cable. (© *Cable & Wireless Communications 2012, by kind permission of Porthcurno Telegraph Museum*)

2 Ships of the squadron, 1858. (© *Cable & Wireless Communications 2012, by kind permission of Porthcurno Telegraph Museum*)

3 The paying-out machinery, 1858. (© *Cable & Wireless Communications 2012, by kind permission of Porthcurno Telegraph Museum*)

4 Departure of the cable squadron from Valentia, 1858. (© *Cable & Wireless Communications 2012, by kind permission of Porthcurno Telegraph Museum*)

5 The Niagara and Agamemnon splicing the cable mid-ocean, 1858.
(© *Cable & Wireless Communications 2012, by kind permission of Porthcurno Telegraph Museum*)

6 Arrival of the Niagara at St John's, Newfoundland, 1858. (© *Cable & Wireless Communications 2012, by kind permission of Porthcurno Telegraph Museum*)

7 The Niagara discharging the shore cable in Trinity Bay, 1858. (© *Cable
& Wireless Communications 2012, by kind permission of Porthcurno
Telegraph Museum*)

8 Hauling the cable ashore in Valentia, 1858. (© *Cable & Wireless Communications 2012, by kind permission of Porthcurno Telegraph Museum*)

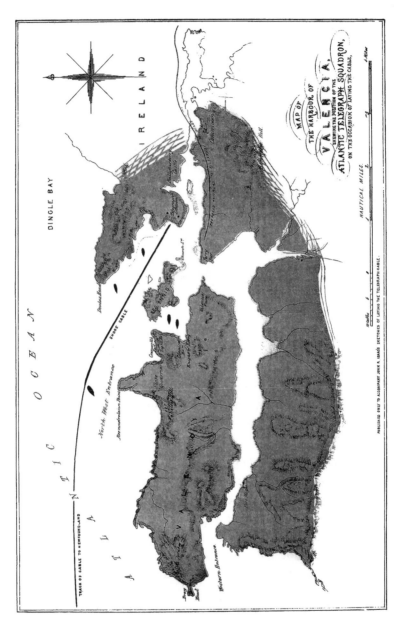

9 The cable's route into Valentia Bay, 1858. (© *Cable & Wireless Communications 2012, by kind permission of Porthcurno Telegraph Museum*)

3

The Stride of

a Full-Grown Giant

Morse and Whitehouse satisfied themselves, and in doing so satisfied their fellow projectors, that the Atlantic cable was entirely viable. After the Newfoundland debacle of 1855, 1856 proved to be a year of achievement and optimism, even to a point of over-confidence. Maury, for one, could see no difficulty at all, writing to Field in May: 'It can be done without fail. It can be done by one steamer, and in case of accident while running, the cable may be recovered. When I say a single ship, of course I mean one large enough to carry the cable and coal. The cable need not be large.' Cyrus Field's view was that submarine telegraphy, while 'in its infancy … is in the act of making the stride of a full-grown giant'.

The Atlantic Telegraph Co. was launched by Field in London in the autumn of 1856. The New York, Newfoundland and London business continued, but in a supporting role only. The idea was that a British company would attract a wider pool of investors, and deal more effectively with governments and cable-makers. The focus of the enterprise thus shifted decisively to London. The new company received an enthusiastic endorsement from *The Times*, whose leader writer had the utmost faith both in the directors and in the engineers involved:

> It is not our custom to come forward as the advocates of joint-stock companies, but surely this project constitutes an exception. The interests of this nation and of the civilised world are so closely bound up with its success that we feel justified in recommending it to the notice of our readers.

It seemed to *The Times* that the scheme could not fail: 'The enterprise must be badly carried out indeed if the revenue … is not sufficient to pay a handsome interest upon the outlay.' As six submarine lines already connected Britain to continental Europe, there seemed no question but that 'what has been so often and so successfully accomplished can surely be as easily accomplished again, when the only additional obstacle is that of greater length, but under more favourable circumstances'. All would fall before the certainties of the scientific age, the unceasing march of technological progress, and the might of Victorian Britain:

> Habit has so familiarised us with the marvellous triumphs of physical science that we have ceased to feel or express astonishment at results, which, not so many years ago, would have been dismissed from the consideration of rational men as the visions of a fantastic imagination.

There was no doubt, no doubt at all, according to *The Times*, that progress would roll on unceasing. 'Many persons now living ... will live to see the year AD 1900. It will be a changed world then!'

This huge confidence was a double-edged sword. On the one hand, it was a vital element in raising money and political support to lay the cable. On the other, the cable's enthusiasts were carried away by their own rhetoric. Self-belief closed down the supporters' critical faculties. At a time when a measured overview of the scheme was called for, the company directors were happy to encourage rising levels of excitement, which made attracting investors so much easier. The scheme, it seemed, could not fail, and would pay large dividends to boot. But of course, the higher the expectations, the greater the disillusion if anything went wrong.

As Morse was completing his London experiments, Field and other directors actively looked for more support. In November, Field was in Liverpool, addressing a crowd of merchants in the underwriters' rooms. The telegraph, he told the assembly, extended in total 42,000 miles beyond St John's, across British North America and the United States. All that was needed to connect it to the European network was 'a single line, 1,900 statute miles in length', which could be laid in less than a week at a cost of £350,000. A sample of cable was passed around. 'Now that's the thing to tell the price of cotton!' exclaimed an admiring broker. Field hardly needed to spell out to this gathering, many of them engaged in trade with America, the value of almost instantaneous information on commodity prices and other commercial news. At the moment, he reminded them, they could not send a letter to New Orleans and receive a reply in less than forty days. With the cable, because of the four or five-hour time difference, their telegram would be in America before they had sent it.

The Atlantic Telegraph Co. was incorporated in October 1856, with 300 shares of £1,000 each. Field took twenty-five shares, as did John Watkins Brett. The capital was increased by £50,000 a few days later. At its launch, the company had £79,000, with a further £75,000 subscribed from America. Before the shares were ever offered publicly, a total of £180,000 had been committed by the promoters and their friends. The £350,000 had been secured by December 1856. This capital was raised without any need to advertise, despite the high individual cost of shares, with many shareholders 'gentlemen of the first standing in this country and the United States'. Brett and Field were to be due an enhanced share of profits, for fifty years following, but this would not come into effect until shareholders received an annual dividend of ten per cent. In exchange for the higher returns, the Newfoundland concessions and other rights and patents held by Brett and Field would be signed over to the company. Further landing rights in Newfoundland and neighbouring territories were also acquired by the new company.

Though it had been easy to raise the money, Field was still keen to attract support from the British and American governments. In October 1856, accompanied by Morse, he gained an interview with the British Foreign Secretary, the Earl of Clarendon. Clarendon was surprised at the scale of the project, and asked what would become of it if the attempt failed. Field is supposed to have replied, 'Charge it to profit and loss, and go to work to lay another.' The minister gave a strong hint that the government would do all it could to help, passing Field over to James Wilson, Secretary to the Treasury and a man of great influence. Wilson entertained Field at his home near Bath. Afterwards he wrote to Field confirming the government's proposals. They would provide ships for new soundings, or to recheck previous soundings, and possibly assist with the job of cable-laying. When the line was working, they would pay £14,000 a year, or four per cent, on the New York & Newfoundland Co.'s capital of £350,000, until profits reached six per cent, and then £10,000 a year for twenty-five years. In exchange, they expected free transmission of official messages, which would take priority over all others. Finally, the British government retained the right to control increases in telegram charges.

The US Senate discussed the cable in January 1857. Some senators were anxious to match the British commitment while reminding the Europeans that 'the whole enterprise has heretofore been conducted with American capital'. As in England, there seemed no doubt that the scheme would succeed. 'On the 4th of July next, if this bill shall pass, there will be, for all practical purposes, an electric girdle round the world.' The bill passed, but very narrowly, and despite some hostility in the House of Representatives, achieved a small majority there too. Government vessels and $70,000 a year for twenty-five years were pledged to the cable company.

The opposition had come from those alarmed that the main cable was to start and finish on territory controlled by Britain. 'Both the termini are in the British dominions. What security are we to have that in time of war we shall have the use of the telegraph as well as the British government?' The question of war had not been raised by the British. Other senators thought the cable would be the means of avoiding wars completely. 'After the telegraphic wire is once laid, there will be no more war between the United States and Great Britain ... I believe that whenever such a connection as this shall be made, we diminish the chances of war.' Another interjected: 'I am willing to vote for this bill as a peace measure, as a commercial measure – but not as a war measure; and when war comes, let us rely on our power and ability to take this end of the wire, and keep it.' Another reminded his audience that American citizens had been the instigators of the enterprise. 'The honor and the glory of the achievement will be due to American genius and American daring. Why should we be actuated by so illiberal a spirit as to refuse the use of one of our steamships when it does not cost one farthing to the Treasury of the United States?'

Cyrus Field's brother Henry wrote later that the chief objection to the scheme was that it involved England:

> The real animus of the opposition was national jealousy – a fear lest they should be giving some advantage to Great Britain. The mention of the name of England has the same effect on them as a red rag waved before the eyes of a mad bull. No matter what the subject of the proposed co-operation, even if it were purely a scientific expedition, they were sure England was going to profit by it to our injury. So now there were those who felt that in this submarine cable England was literally crawling under the sea to get some advantage of the United States.

But any idea of taking the line directly to the United States was put to rest by the Atlantic surveys of Lieutenant Maury and Captain Berryman. When Field had visited Washington in June 1856, he secured the use of Berryman and the steamship *Arctic*. Berryman took further soundings every thirty to fifty miles along the route, and also used scoops or quills to bring up specimens from the bottom, all of which proved to be a satisfactory mix of fine sand and powdered shell. He wrote to Field in November that year that a direct route from Europe to the United States would be 'quite useless, if not impracticable'. Berryman's views were swayed by a conviction that transmitting signals over 3,000 miles could prove to be impossible. He suggested also that the current and deep waters of the Gulf Stream would pose a problem, though the area had not been fully explored. All of this influenced him in favour of the route to Newfoundland, where the only danger anticipated was iceberg damage.

This had not satisfied all the American politicians. C.C. Chaffee of the US House of Representatives wrote to Lieutenant Maury in December 1856: 'Is there a point, *under our flag*, which would answer for the western terminus?' Maury attempted to discourage him: 'The difficulties are manifold, and in the present state of the telegraphic art, they may be insuperable.' He too believed that 3,000 miles was an impossible length to transmit an electric current, whereas 1,500 (nautical) miles was 'electrically practicable'. Although there was another possible route from the United States to Europe, with a maximum unbroken length of cable of only 1,500 miles, this passed through the deepest water and across volcanic regions. It also needed a relay station in the Azores, where the cable could fall under the control of the Portuguese government.

Any idea that a line could bypass British North America and go 'direct to the shores of the United States' constituted a threat to the Atlantic Telegraph Co., whose resources were by then committed to the Newfoundland route. Field felt that he must fight any alternative scheme, and did so by publishing his letters from Maury and Berryman. It was rumoured that another company was being formed to project a cable direct to the United States. Field claimed not to believe these stories – for he did not think 'that the capitalists in London or elsewhere can be found to support such a scheme', which was for a cable twice the length of Field's. Furthermore, he knew that no survey had been carried out. Yet he took the idea seriously enough to launch a preemptive strike against it. Field, the American who loved England, believed that the venture must be a joint one, and that its completion would reaffirm the strength of alliance between the old country and the New World. A joint enterprise would 'serve as a guarantee to the world that in the case of war, that the cord is never to be broken'. He was adamant that his Liberal friends in the United Kingdom would not undermine United States interests. 'The British government interfere with the free use of the cable, even in war! The spirit of the age is against such an act.'

Field's global vision, which had first drawn him to the telegraph, was undiminished. Though he was certainly interested in promoting Anglo-American relations and trade, he saw a wider picture. He had been impressed by an international maritime conference in Brussels in 1853: 'This research was not undertaken for the exclusive advantage of any one person or nation, but for the benefit of commerce, the advancement of science, and for the benefit and improvement of the whole human family.' He also argued that it would be political folly for the United States not to be involved in the cable:

But suppose we should stand aloof, and that the enterprise now on foot should be abandoned by our citizens and government, and then suppose war to come; in less than six months after its declaration, the British government could, on its own account, have a wire stretched along this telegraphic plateau between Newfoundland and Ireland.

Morse's underground experiments seem to have satisfied Field entirely. If he had any doubts, he did not voice them. He reminded the public of Brett's Channel cable of 1851. 'The result of this decisive experiment, favourable alike in its national, commercial, social, and, though last not least, in its remunerative aspects, has been such as to disarm all prejudice, and to encourage a desire for the utmost possible extension of similar undertakings.' Again he repeated that America had missed out, while six cables already connected England to continental countries.

In November 1856 Field was able to announce a timetable, with the cable-laying planned for summer 1857, and not later than spring 1858. Two steam-ships would be used, starting in the middle of the Atlantic and taking about eight days. Once complete, the cable 'will constitute the chief medium through which all the important business transactions between the Old and New World will be effected. The transmission of intelligence for the press in both continents will also form a most important feature of its usefulness.' He argued that the capital involved was small, 'compared with the magnitude and the national importance of the work'. As running expenses were very low, 'it appears difficult to over-estimate the commercial returns that will accrue from this undertaking'. Even a very few commercial telegrams each day, with no other source of revenue, would produce a large return on the capital. He envisaged busy traffic between 10 a.m. and 3 p.m. from Britain, with messages arriving in the United States in time for the opening of business.

The board of directors accepted Whitehouse's scheme that the cable core should be slightly larger than first suggested. Its seven copper strands would be insulated with a treble layer of gutta percha, so that it measured three-eighths of an inch in diameter. The core, surrounded by jute yarn saturated with tar, pitch, boiled oil and common beeswax, would be made by the Gutta Percha Co. The armouring consisted of eighteen strands of seven wires each of charcoal iron bright wire, making a total weight of one ton to the mile. The breaking strain was 65 hundredweight, over three tons. It was calculated that the cable was double the strength required to support its own weight while being laid in the deepest water on its route, which was 2,400 fathoms. As there was no factory with the capacity to armour the whole cable, the 2,200 miles of line, to be completed by the end of June 1857, were split between the two leading submarine cable-makers, R.S. Newall of Birkenhead, and Kuper & Co. of Greenwich, which became Glass, Elliot & Co. during the course of the contract. These companies, both originally manufacturers of pit ropes, had developed great expertise and an even greater rivalry in this new field.

The Atlantic company's board of directors, confirmed in December 1856, included the London bankers Thomas Baring and Samuel Gurney, provincial merchants from Liverpool, Glasgow and Manchester, and the engineers William Fothergill Cooke, John Brett, Professor William Thomson and Charles Bright. The board did not include

Field, Morse or Cooper, for directors must be resident in Britain. Instead, they and Wilson G. Hunt (a New York merchant), Hiram O. Alden (vice president of the American Telegraph Co.), Watts Sherman (a New York banker), Alex H. Rice (merchant and mayor of Boston), and Lieutenant Maury were made 'honorary' American directors, joined by the Hon. John Ross, the Hon. John Young and the Hon. George E. Cartier representing the Canadian provinces.

The move to London also effectively ended any role for Gisborne in the project. Although increasingly marginalised, he had continued as chief engineer to the New York, Newfoundland & London Electric Telegraph Co. In January 1857 his term ended and his contract was not renewed. This was in spite of his pleas that he was responsible for tying up privileges which 'preclude the possibility of opposition or competition in transatlantic telegraphing', and for bringing Newfoundland into a fifty-year contract with 'some of the wealthiest and most eminent men in Europe and America'.

Charles Bright had officially become chief engineer to the Atlantic Telegraph Co. at the end of 1856, at a salary of £1,000 a year, and Whitehouse was appointed electrician on the same terms. The huge public interest in the cable attracted dozens of enquiries from people seeking jobs, as engineers and at sea, and also offering technical and geographical suggestions. Newspaper letters columns were also packed with ill-founded advice for the engineers. The idea that the cable could be buoyed just below the ocean's surface, as Hubbell had proposed, was a popular one. Someone describing himself as a naval expert suggested that nothing more than a handspike was needed to control the laying of the line. Others put forward the idea that the cable should be manufactured on board ship and submerged immediately, so that no joints were needed. Albert, the Prince Consort, wanted the whole cable protected by glass tube, and had to be discreetly rebuffed.

Field was back in the United States and the British provinces of North America during February and March of 1857. The US steam frigates *Niagara* and *Mississippi* had been ordered to assist with the telegraph. In Nova Scotia, the House of Assembly granted the new company exclusive rights, and other legislation was in progress in the various jurisdictions. Realising that he could not cover all these areas personally, Field proposed that Edward Mortimer Archibald, of Halifax, Nova Scotia, a barrister and former attorney general of Newfoundland, should represent the company in the North American colonies, at a salary of £500 plus expenses. He put this for the board's approval on his return to England in April.

In London, Field discovered that the Gutta Percha Co. had almost completed the cable core, and that 810 miles of cable were fully armoured. By mid-May, 860 miles lay finished at Birkenhead, and 780 miles at Glass & Elliot's works in Greenwich. Charles Bright had the design and manufacture of the paying-out machinery in hand. The power for the Atlantic telegraph was to be generated by a 'giant voltaic battery of ten capacious cells'. Its plates of platinised silver and zinc offered a surface area of over 2,000 square inches. If the zinc were periodically replaced, the battery would be 'exhaustless and permanent'. This battery was designed by Whitehouse, modified from a type already used extensively in telegraphy. Siemens & Halske of Berlin were asked to quote prices for Morse instruments which printed signals as dots and dashes on rolls of paper. Samuel Canning, who had been in charge of cable-laying throughout the events off

Newfoundland, was appointed engineer responsible for the cable on board ship.

Yet there were still anxieties about the cable itself. Bright had argued from the time of his appointment that the core and insulator needed to be larger. He was overruled by the businessmen, anxious to press ahead and complete the crossing the following summer. Morse, after all, supported their case, believing that a larger cable would be a worse conductor. Bright and Whitehouse were particularly concerned about Newall's part of the cable in Birkenhead, and its 'deficient insulation'. They were allowed to experiment on it in the factory, on condition that Professor Thomson was consulted and involved. This was not the only friction with Newall, an acerbic and litigious character, who was in dispute over a price

Glass, Elliot & Co.'s cable works at East Greenwich.
(*Cable & Wireless Archives, Porthcurno*)

for shore-end cable which the company considered too high but he declined to reduce.

Worse was to come. At some point, as the cable neared completion, it was discovered that the Birkenhead and Greenwich sections had been made with threads running in different directions. This mistake occurred because a hand-made specimen from Glass, Elliot & Co., 'the lay of which happened to be opposite to that of their machine-made ropes' had been shown to Newall. The Birkenhead contract was then based on this sample. Somehow, over the following months, the blame for this mistake became – quite unfairly – attached to Newall. The Atlantic company was growing increasingly close to Newall's competitor, Glass & Elliot, while Newall did little to endear himself personally to his clients.

Whatever the truth behind the mix-up, Bright always denied any responsibility. The confusion had arisen during the autumn of 1856, while he and Whitehouse were preoccupied with the signalling trials with Morse, and before he was appointed Atlantic company engineer. In any case, the problem proved quite straightforward to remedy – the two parts of the cable were easily joined. Bright was sure afterwards that 'undue importance has been ascribed to this difference of lay in the two lengths of cable'. But the conflicting threads were a simple matter for the public to understand and latch on to, far easier to comprehend than the electrical obstacles which baffled even the leading electricians. So the thread issue continued to resound, tainting Newall's reputation further, and was later used to illustrate the company's supposed incompetence.

Nor had the cable-handling problems been resolved. Bright and Canning calculated that *Niagara*, a 'splendid new frigate', a screw-corvette said to be the largest and finest ship in the US Navy, did not have the capacity to take her section of the line, so that a third would have to be stowed on the quarter deck. This posed a danger to the men paying out the cable. The Atlantic board enquired after the screw steamer *City of Baltimore* as an alternative. The *Baltimore* was available, but even with the cost of hire reduced to £2 7s 3d per ton per month, it was considered too expensive. HMS *St Jeanne d'Arc* was offered; she could house the section of cable but would not fit in Newall's dock. There were in any case political reasons to stick with the *Niagara*. The frigate had arrived in London with a full complement of naval officers and a crew of 500, which certainly could not be accommodated along with her part of the cable. A temporary re-fit in the royal dockyards was decided, so *Niagara* set out for Portsmouth, to be adapted to take 'her long passenger' from Newall.

As work neared completion, Glass & Elliot's yard was hon-oured with a visit by the Lord Mayor of London, transported there in 'a corporation state barge, rich enough in flags and bunting to have satisfied the Queen of Sheba'. His party contrasted oddly with their surroundings, for the Greenwich factory was not a pleasant place at low water. 'The shore is stony, and unsavoury in hot weather, when all the abominable flotsam and jetsam of the dirtiest river in Europe evaporate enough miasma to poison an alligator.' Land was reached by way of a narrow plank, which:

> necessitated feats on the part of stout gentlemen which in
> a circus would have covered them with immortal honour.
> Plank-walking under such circumstances is always difficult,
> and when the performance has to be gone through by a

portly functionary in full Court dress and cocked hat, anxious to preserve his dignity, yet afraid of losing his balance, the result has no medium, but is at once either sublime or ridiculous.

Of the cable, 900 miles had been completed when the Lord Mayor called, although work was behind schedule as fine wire was being used at a rate faster than all the wiredrawers in the country could supply it. The total length of copper and iron wire in the cable's core and armour was 340,000 miles, enough to travel thirteen times round the world, more than enough to reach the moon. The ship which would lay the Greenwich section of cable, the *Agamemnon*, a Royal Navy screw steam ship of 60hp which had seen service as an admiral's flagship during the bombardment of Sebastopol in the Crimea, had not yet arrived. Her mooring of ten anchors was being prepared, designed to stop any motion while the 'ponderous coils' were transferred into the hold. The cable would be carried across supports fixed on ten barges between factory and ship, wound slowly by a 12hp engine into one coil, 45ft in diameter and eventually 12ft high.

The completion of Newall's half of the cable was celebrated in mid-June with a banquet at the Birkenhead factory. The cable, 1,250 miles of it, had been coiled on board barges alongside the warehouses of the Birkenhead Float, awaiting the *Niagara*. It had been settled without a doubt, a day or two earlier, that a current of electricity would indeed pass the entire length of the cable. Newall's hundreds of workmen were joined at the celebration meal by some of the Atlantic company directors, as well as local worthies. Toasts were raised to the President and people of the United

States as well as to the Queen, and hopes expressed that the telegraph 'would nip in the bud those bitter feelings which unfortunately had occasionally arisen between the mother country and her descendants'. Memories were fresh of the bloody and damaging Crimean War, for which Newall had made the submarine cable. Many were conscious that better transatlantic communications at that time might have brought the United States to the aid of Britain and France. As it was, the Russians had managed to divert the United States from the European conflict by engaging her attention in another direction, offering negotiations for the sale of Alaska at cut price.

The peace which the Birkenhead speakers anticipated so keenly was Pax Britannica. The cable, which the Americans still saw as their idea and their venture, was undergoing a hijack by British interests. As the original Newfoundland to New York telegraph had grown into a transatlantic line, so now all the talk was of encircling the entire globe. Newall's men were about to start work on a new Mediterranean cable, part of a projected line to India, and the further possibilities began to carry away some of the diners at the banquet. 'This Atlantic submarine telegraph, large an undertaking as it is, is only part of a vast system which is spreading itself in every direction.' The Indian cable, it was predicted, 'will very soon be extended to China, to Australia, across the Pacific Ocean to California; and this country will form the centre of a system of telegraphs, whose wires will spread over the face of every great country, and under the surface of every great sea'. And here was the rub: 'Thus, in future, when the great heart of England throbs, its pulsations will instantly reverberate to the uttermost parts of the earth.'

It emerged, though, that the new Atlantic Telegraph Co. did not have the powers needed to take over the Newfoundland concessions and operate the telegraph away from British shores. It needed a new corporate identity. There was no precedent for this situation. The company was therefore set up and regulated by an Act of Parliament, which placed it under limited liability rules but allowed it to run the line between Britain, Ireland and Newfoundland. The government was also firm that the company's offices must be in London.

Many shareholders were angered by this, not least because decisions were being taken out of their hands. Some argued that it would 'virtually place the perpetual management of officers in the hands of directors who were resident in London'. But by the time a shareholders' meeting was called in June 1857, the Bill was already passing through Parliament. Sir William Brown, the Liverpool merchant and Member of Parliament who chaired the company, explained that the government had threatened to negotiate with another company if the Atlantic directors would not agree. Brown reminded the shareholders that four government vessels were being used, saving an estimated £100,000. The *Agamemnon* had been converted for cable-laying and was about to start taking the cable. There was also the promised £14,000-a-year subsidy. If the bill were delayed, there could be no attempt to lay the cable before the August gales, effectively the end of the laying season, 'retarding operations for a considerable period'.

Shareholders were also concerned that the government was insisting on a right to veto directors. This power was needed so that no party could take control of the company by buying up shares. The company's solicitor, Mr Freshfield, admitted that 'such an amount of government interference was in principle objectionable'. But shareholders had little choice but to meet the government's demands, or 'they would virtually ruin their undertaking'. The Act to incorporate the company was passed in July 1857.

Instrument clerks had already been appointed at £110 and £100 per annum. Sixty men were to be engaged for paying out the cable. In the last week of June, the *Niagara*, adapted to take the cable in four coils, had arrived at Birkenhead. Coiling on board began, at a rate of forty-eight to fifty miles in twenty-four hours. *Agamemnon*, meanwhile, was in Greenwich, receiving the cable in one coil. She had been 'jury rigged', that is, had her heavy masts and rigging replaced by lighter spars and ropes so that she was steadier and easier to manage under steam. The ship was of an unusual design, her engines at the stern, and amidships a vast hold, 45ft square and 20ft deep, capable of holding the cable in a single coil. By mid-July, 1,100 miles were on board the *Agamemnon* and 900 miles on the *Niagara*. The collaboration between two of the largest warships in the world was seen as emblematic of a great step forward in Anglo-American relations. In New York, *Harper's Journal* commented on the way that swords were being beaten to ploughshares: 'What would Nelson and Collingwood have said of meeting a foreign first-rate in mid-ocean *to lay a cable* at the bottom of the sea!'

Once the *Agamemnon* had received her section of cable, the ship was set out with tables for a great celebration, to which her crew and the Glass & Elliot workmen and their families were invited. Each man was given three pints of beer, used for an increasingly extravagant succession of toasts. As the vessel left the Thames, a salute was fired, and thousands lined the riverbanks at Greenwich.

Field returned from America by the end of July, with favourable news, including a request from the New York Associated Press to enter an agreement with the company to transmit 'intelligence'. HMS *Cyclops* was in the Atlantic, checking Berryman's soundings. Maury, meanwhile, had been considering the optimum time for cable-laying by studying past weather patterns in the ocean. During the summer months, there were never gales on the western part of the route, and few or none on the eastern section, except near the coast of Ireland. In fact, meteorological records from both sides of the Atlantic showed that there had been no major storms in the preceding fifty years during June or July. The main problem in the west was fog, especially in June. Cable ships, ironically, had no way of communicating with each other except by semaphore signalling, so fog was potentially a great handicap. Fog was never a problem in winter, but any attempt at that time faced dangers from ice. August was the best month to avoid icebergs. Maury concluded that, all things considered, the weeks between 20 July and 10 August offered the best prospects for the project. So the ships were to be despatched to be ready to start work on 20 July, and if all went to plan the job could be finished in two to three weeks. In case the weather did not measure up to Maury's predictions, there was a contingency plan. Ships in storms would often have to slip their anchor, and it was envisaged that in an Atlantic gale the cable-layers may have to slip the telegraph line itself in order to save their vessels. In that case, a special line of iron wire would be spliced to the cable, and buoyed so that the cable itself rested safely on the seabed until the storm passed. Afterwards the cable could easily be retrieved and rejoined.

While Maury planned the laying, the directors published a paper setting out their own vision of the cable's impact upon the world. They assumed that messages would pass at least fourteen days faster than could possibly be achieved by any other means. They dismissed the idea that the cable could be used for any warlike purpose, and in fact Morse wrote that 'the chances of conflict and misunderstanding must be diminished in an incalculable degree' for 'all wars arise in ignorance and misunderstanding of the real objects and interests of the races by which they are waged'. New York would become a suburb of London, and Washington the western half of Westminster. The cable, said the company, would do more for international peace than the spending of ten millions a year on each side of the Atlantic on 'armed Leviathans'. Communication was to resolve all problems.

4

Lightnings through Deep Waters

On the last day of July, the two lengths of cable met for the first time, in the harbour of Queenstown, now called Cóbh, on the south coast of Ireland. They were joined together and messages were sent successfully through the whole distance, 2,500 miles, in less than a second. 'Everything works beautifully,' reported *The Times*. Some transfer of shore-end cable was required from the *Agamemnon* to the *Niagara*, as a late decision had been taken to start laying the cable in Ireland rather than mid-ocean. Bright strongly favoured the idea of starting in mid-Atlantic, as the ships could wait for good weather and then complete laying in half the time it would otherwise take. But the electricians preferred to lay from shore, as they could be in touch with land throughout the process. This appealed to some of the directors, struck by the novelty of being able to talk from London to a vessel steaming across the ocean, and Bright was overruled. So on 3 August, the 'wire squadron', which included HM Sounding Vessel *Cyclops*, HMS *Leopard*, and the US paddle frigate *Susquehanna*, steamed out of Queenstown for the island of Valentia, Europe's most westerly port.

John Lecky, a boy in Valentia, later recalled how the Atlantic cable transformed his quiet home:

> In 1857 when the ships arrived we had so many visitors in the island it has always been a wonder where they got accommodation. To celebrate the laying of the cable, the Knight of Kerry gave a banquet and dance and these were held in John Driscoll's store in the slate yard. Most fortunately there were sufficient slabs ready to lay down for the floor and others for the dinner table.

This event was described in *The Times* as 'an elegant *déjeuner*'. Amidst the odd surroundings, spirits were high. The Lord Lieutenant of Ireland summed up the general view:

> We are about to establish a new material link between the old world and the new. Moral links there have been – links of race, links of commerce, links of friendship, links of literature, links of glory – but this, our new link, instead of superseding and supplanting the old ones, is to give a life and an intensity which they never had before.

The cable was brought ashore, and signals successfully sent from the beach through the full length of the *Agamemnon*'s cable. The following day, the fleet set off on its course of 1,834 miles to Trinity Bay, Newfoundland. Valentia was the closest point in the British Isles to the New World, and her sandy

Sketches of the *Cyclops* (top) and the *Niagara* by E.W. Cooke.
(*Institution of Engineering and Technology*)

coves with deep water provided ideal conditions in which to land a cable. The ships would carry a surplus of 600 miles of line so that even if there were deviation from the planned course, there was no danger that it would run out. Although Newall was still raging in the letters columns of *The Times* for being blamed for 'a most egregious blunder', the mistake over the cable's thread was not a fatal one. In the event the two sections were spliced without much difficulty. Charles Bright, the engineer in charge of the expedition, ensured that

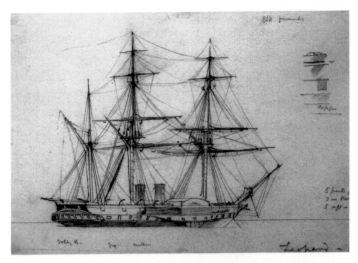

Sketch of the *Leopard* by E.W. Cooke.
(*Institution of Engineering and Technology*)

sections of cable from both suppliers were laid at an early stage, to prove that they could easily be joined at sea, and to show that the misunderstanding had not caused huge difficulty. All seemed on course for success.

The expedition left Valentia on 7 August, laying shore-end cable from the *Niagara*, following the course of the *Cyclops*, which was taking soundings. The *Agamemnon* was to follow, and would take over the laying in mid-ocean once the *Niagara's* cable was spent. Eight miles out, the shore end was spliced to the main cable. By noon the following day, forty miles had been laid. Four hours later, as the water deepened, a brake had to be placed on the cable so that the speed at which it left the ship was roughly the same as the forward movement of the vessel. Until that point, the laying had been regulated by the action of the machinery against the ship's forward motion. The brake, made by friction drums against paying-out sheaves, was gradually increased through the night, depending on the speed of the *Niagara* and the depth reported by the *Cyclops*. The following evening, when about 180 miles had been laid, the cautious Bright, happy that the men and machinery were 'well at work', allowed the ship's speed to increase from three or four knots, to five. Early in the morning of 10 August, the ocean's depth increased rapidly, from 550 fathoms to 1,750 in a distance of eight miles. 'Up to this time', reported Bright:

> seven hundredweight strain sufficed to keep the rate of the cable near enough to that of the ship, but as the water deepened the proportionate speed of the cable advanced, and it was necessary to augment the pressure by degrees, until, in the depth of 1,700 fathoms, the indicator showed a strain of 15 hundredweight [three quarters of a ton] while the cable and ship were running five and a half and five knots respectively.

By noon, with the vessel 214 miles from shore, and having laid 255 miles of cable, there was an increasing swell and a strong breeze. Morse, who held a watching brief for the US government, 'had to retire to his berth as soon as the elements asserted themselves, and was scarcely visible again till all was over'.

After this, with the depth approaching 2,000 fathoms, about two miles, the strain had to be increased to a ton. 'At six in the evening', said Bright, 'some difficulty arose through the cable getting out of the sheaves of the paying-out machine, owing to the tar and pitch hardening in the grooves, and a splice of large dimensions passing over them.' Bright managed to retrieve the situation by fitting extra guards to the machine, and removing the congealed tar with oil. While this work was in progress, laying had to be halted, the cable temporarily held by stoppers until it could be re-attached to its pulleys. Bright considered this to be a significant achievement, 'showing that it is possible to lay in deep waters without continuing to lay out cable – a point upon which doubts have been frequently expressed'. This had been an anxiety beforehand, as the 'vast strength and grip of machinery' needed to support the weight of cable at depth had caused problems in the Mediterranean.

But then the cable started to run away at a speed much faster than that of the ship. The vessel had slowed to three knots, but the cable was leaving at five and a half or more. The wind and sea meanwhile increased their force, and the task was complicated by a current which carried the cable at an angle from the ship's course. The brake, or retarding force, had to be increased to 25, then 30, and finally 35 hundredweight, or one and three quarter tons, as the cable issued forth at a speed 'more than it would have been prudent to permit'.

The cable's progress was slowed to five knots, when in the early hours of 11 August, with 335 miles laid and in 2,000 fathoms, it snapped. Bright, who had been attending to the brakes himself, was sure that matters were under control and briefly left the scene to check the speed of the ship and to see how the cable was emerging from the hold. In the moments he was away, the machine was left in the charge of an experienced mechanic, one who 'had been from the first engaged in its construction and fitting'. This man, though, failed to release the brakes to compensate for the ship's descent in the swell. Bright was moving forward on the ship when he heard the machine stop. 'I immediately called out to ease the brake and reverse the engine of the ship, but when I reached the spot the cable was broken.' He later blamed the exhaustion of his men for the mistake:

> When the rate of the wheels grew slower as the ship dropped her stern in the swell, the brake should have been eased; this had been done regularly before whenever an unusually sudden descent of the ship temporarily withdrew the pressure from the cable in the sea, but owing to our entering the deep water the previous morning, and having all hands ready for any emergency that might occur there, the chief part of my staff had been compelled to give in at night through sheer exhaustion, and hence been short-handed. I was obliged for the time to leave the machine, without, as it proves, sufficient intelligence to control it.

Bright would not criticise the machinery itself. 'It has been suggested as a cause of the failure that the machinery is too massive and ponderous; my experience of its action teaches otherwise. For three days in shallow and deep water as well as in rapid transition from one to the other nothing could be more perfect than its working'. He argued that, since the machinery had worked well in shallower water 'where the weight of the cable had less ability to overcome its friction and resistance', it could not be too heavy for deep water where extra friction had to be introduced to restrain the cable's rapid passage from the ship. He suggested some slight improvements to the brakes, though expressing complete confidence in the machinery as it stood. 'It must be remembered as a test of the work which it has done, that unfortunate as this termination to the expedition is, the longest length of cable ever laid has been paid out by it, and that in the deepest water yet passed over.'

The newspapers were told on 12 August that 'an accident of some description' had occurred. Wildman Whitehouse, who had stayed on shore as his health was not good, was in the temporary telegraph station set up in a hut and tents on Valentia beach. He received constant signals and messages until the cable's failure, and could tell from his tests that the problem was 350 to 400 miles off shore. 'In the absence of any means of communicating with the ships, it is of course impossible even to guess at the cause of this embarrassment.' But there were hopes that the cable could be hauled in and repaired. 'With a view to this possibility, signals of great energy are now being constantly sent from this station through the cable, but no answer has at present been received.' Two days later, the *Cyclops* steamed into Valentia to confirm the worst, carrying letters from Field announcing that cable-laying had been abandoned, at least for a time. The season was late, there was little hope of recovering the lost cable, and there was not enough line remaining to complete the distance without it.

So only two weeks after the flotilla had set out with such confidence, the Atlantic directors faced the ignominy of an approach from the Red Sea Co., which wanted to buy the unused cable. This offer was considered, although Bright believed the lost cable could be retrieved at a relatively low cost and the whole re-laid the following summer.

Despite the considerable setback, the young engineer remained upbeat:

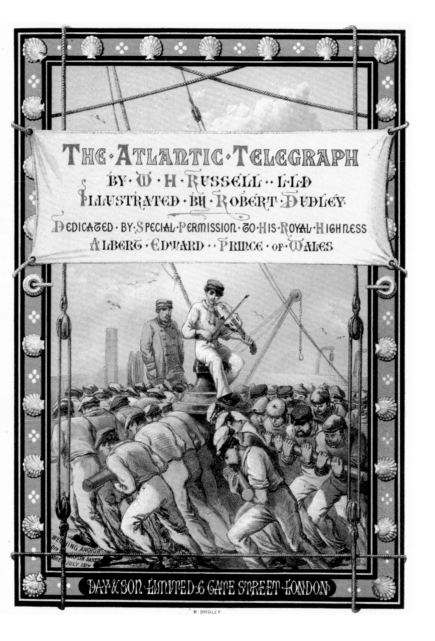

1 Frontispiece by Robert Dudley from W.H. Russell's *The Atlantic Telegraph*
(Day & Son Ltd: London, 1865).

Coating the electric core with jute.

Covering the cable with jute and tar.

Covering the core with wire.

Coiling down the cable in the tank.

3 *(opposite above)* Building the land line from Dublin to Valentia.
(Cable & Wireless Archives, Porthcurno)

4 *(opposite below)* Telegraph station in a tent at Valentia, drawn by Robert Dudley.
(Institution of Engineering and Technology)

5 *(above)* The crew of the *Agamemnon*, 1858.
(Institution of Engineering and Technology)

6 Reels of gutta percha-covered wire are placed in tanks
at the Greenwich works.

7 Valentia in 1857-58.

8 Trinity Bay, Newfoundland:
the telegraph house in 1857-58.

9 The Trinity Bay telegraph house mess room, 1858.

10 (*above*) An unforeseen hazard:
a whale crosses the *Agamemnon*'s cable, 1858.

11 *(opposite above)* Making steel wires for the 1865 cable.
(*Cable & Wireless Archives, Porthcurno*)

12 *(opposite below)* Coiling the cable in the Greenwich works.

13 *(above)* The *Great Eastern* at Milford Haven.
(Cable & Wireless Archives, Porthcurno)

14 *(opposite)* Samples of cable and signals, 1858 and 1866.
(Cable & Wireless Archives, Porthcurno)

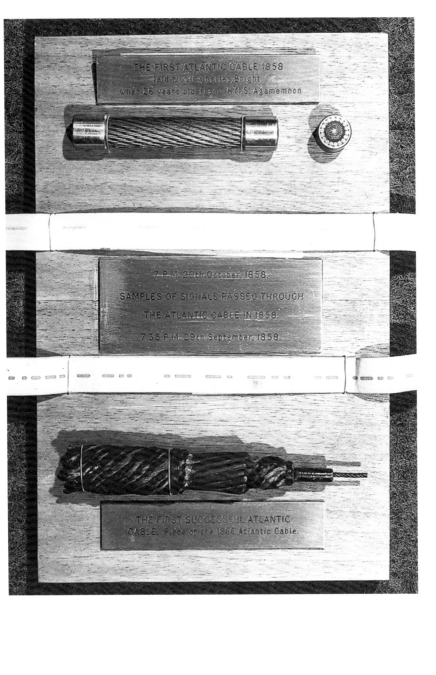

THE FIRST ATLANTIC CABLE 1858
laid by Sir Charles Bright,
when 26 years old, from H.M.S. Agamemnon

7 P.M. 20th October, 1858.
SAMPLES OF SIGNALS PASSED THROUGH
THE ATLANTIC CABLE IN 1858.
7.35 P.M. 29th September, 1858.

THE FIRST SUCCESSFUL ATLANTIC
CABLE. Piece of the 1866 Atlantic Cable.

15 *(opposite above)* Cable passing from Glass & Elliot's works
into a hulk on the Thames.

16 *(opposite below)* Paying-out machinery.

17 *(above)* Cable being loaded from an old frigate on to
the *Great Eastern* at Sheerness.

18 Visit of the Prince of Wales to see cable coiled aboard
the *Great Eastern*, 24 May 1865.

19 The *Great Eastern* cable-laying machinery, *c.*1865.
(*Cable & Wireless Archives, Porthcurno*)

20 Foilhummerum Bay, looking seaward.

21 Foilhummerum Bay, where the cable came ashore in Ireland, 1865.

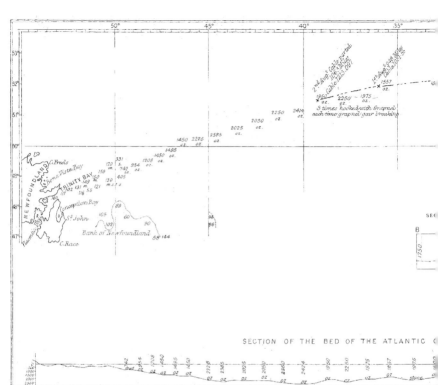

ATLANTIC TELEGRAPH CABLE 1865.

Chart

SHEWING THE TRACK OF

THE STEAM SHIP "GREAT EASTERN" ON HER VOYAGE FROM VALENTIA TO NEWFOUNDLAND
WITH THE SOUNDINGS _ THE DAILY LATITUDE AND LONGITUDE _ THE DISTANCE RUN
AND THE NUMBER OF MILES OF CABLE PAID OUT

LONDON_ DAY & SON (LIMITED)

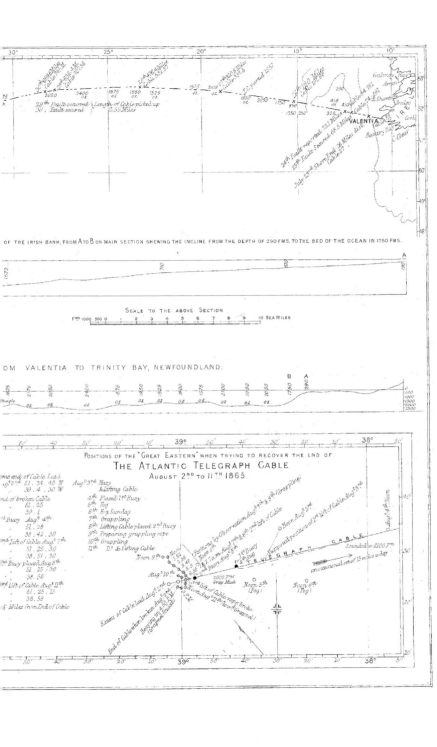

OF THE IRISH BANK, FROM A TO B ON MAIN SECTION SHEWING THE INCLINE FROM THE DEPTH OF 290 FMS. TO THE BED OF THE OCEAN IN 1750 FMS.

SCALE TO THE ABOVE SECTION

F^{MS} 1000 500 0   1   2   3   4   5   6   7   8   9   10 SEA MILES

OM VALENTIA TO TRINITY BAY, NEWFOUNDLAND.

POSITIONS OF THE "GREAT EASTERN" WHEN TRYING TO RECOVER THE END OF

THE ATLANTIC TELEGRAPH CABLE

AUGUST 2ND TO 11TH 1865.

22 (*previous page*) The course of the *Great Eastern*.

23 *(opposite above)* Foilhummerum Bay from Cromwell Point fort:
the *Caroline* and boats laying the earth-wire, 1865.

24 *(opposite below)* The *Caroline* laying the shore end cable,
Port Magee, 22 July 1865.

25 *(above)* The *Great Eastern* under way, 23 July 1865.

26 *(overleaf)* Cable passing out of a tank on board the *Great Eastern*.

27 Splicing the cable after the first accident, 25 July 1865.

28 Searching for a fault in the recovered cable, 31 July 1865.

29 *(opposite)* View from the paddle-box of the *Great Eastern*
en route to Newfoundland.

30 (*above*) Getting out one of the large buoys for launching,
2 August 1865.

31 (*opposite above*) In the bows, preparing to grapple for the lost cable,
2 August 1865.

32 (*opposite below*) The forge on deck, preparing a capstan.

33 (*overleaf*) Forward deck cleared for the final attempt at grappling,
11 August 1865.

34 (*above*) Buoys and grapnels used to recover the 1865 cable.
(*Illustrated London News*)

35 (*opposite*) Marking the spot where the cable was lost in 1865.

36 The cable arriving in Heart's Content Bay, 1866.
(*Cable & Wireless Archives, Porthcurno*)

37 Landing the cable in Heart's Content Bay, 1866.
(*Cable & Wireless Archives, Porthcurno*)

38 Officers of the *Great Eastern*, photographed in 1869, as they set out on another cable-laying voyage. (*Cable & Wireless Archives, Porthcurno*)

39 The cliffs at Porthcurno Bay, Cornwall.
(*Cable & Wireless Archives, Porthcurno*)

40 Laying the shore end of the French Atlantic telegraph
in Brittany, 1869. (*Cable & Wireless Archives, Porthcurno*)

41 The French Atlantic telegraph expedition, 1869. (*Cable & Wireless Archives, Porthcurno*)

I do not perceive, in our present position, any reason for discouragement; but I have, on the contrary, a greater confidence than ever in the undertaking. It has been proved beyond a doubt that no obstacle exists to prevent our ultimate success. I see clearly how every difficulty which has presented itself in this voyage can be effectively dealt with in the next. The cable has been laid at the expected rate in the great depth; its electrical working through the entire length has been most satisfactorily accomplished, while the portion laid actually improved in efficiency by being submerged, from the low temperature of the water and the close compression of the texture of the gutta percha.

In Bright's opinion, the structure of the cable was 'expressly adapted to our requirements'. Its weight in the water was right for the depth, he thought, so that the stress on it was manageable; a lighter cable would take longer to lay, so in the laying would be exposed to greater strain.

The inquest began on 20 August. The naval officers generally concurred with Bright about the cause of the mishap, and gave their opinion that a further attempt to lay the cable that year, in October, would be too dangerous. They also insisted that a future expedition would best start mid-ocean. Bright asked that there be more men, as the 'wearing and anxious nature of the work' called for 'three separate relays of staff, and to employ for attention to the brakes a high degree of mechanical skill'. A scientific committee, consisting of William E. Everett of the US Navy, chief engineer of the *Niagara*, along with Mr Lloyd, chief of the steam department of Her Majesty's Navy, and two eminent marine engineers, William Penn of Greenwich and Joshua Field of Maudslay,

Son & Field of Lambeth, was despatched to Plymouth to examine machinery and mechanical appliances. It seemed that, in the rush to adapt the *Niagara*, there had not been sufficient attention to the design of suitable paying-out machinery. The committee identified ways in which the machines might be improved, and asked Joseph Whitworth, the leading mechanical engineer of his day, to consider the suggestions. Whitworth looked, but declined to make any changes. William Thomson had also written to his fellow directors arguing for improvements to the purity of the copper in the cable's core, and also complaining that the appliances used for hauling in the cable were inefficient.

Despite some criticism of the Atlantic company in the columns of *The Times*, confidence in the cable was not seriously dented. Field, who had been aboard the *Agamemnon* when the cable snapped, rushed back to Portsmouth on the *Leopard* and tried to organise another expedition immediately, but was overruled by his fellow directors. By early September, a new attempt was being planned for the following summer. Professor Thomson was to oversee a competition for the best telegraphic instrument. He also chaired a committee to enquire into the electrical problems which had manifested themselves, the report from which was passed to Whitehouse.

In October the *Leipzig* set out to recover the sunken cable. Whitehouse meanwhile was planning tests on unused cable from the *Niagara*, stored with that from the *Agamemnon* in a vacant powder magazine belonging to the Admiralty in Keyham, Plymouth. Both ships had been unloaded at Plymouth docks, which had the best facilities of any port for landing cable. Bright had already tested both cables on board ship after the failure, finding them to be 'in as perfect condition as when leaving the works at Greenwich and Birkenhead'. Yet as William Thomson continued to do tests on them over the winter of 1857–58, he found numerous faults. The insulation was suspect.

When Cyrus Field returned to New York at the end of 1857, he discovered that his own fortune had collapsed in a commercial crisis which gripped the city. His company had been forced to suspend payments. Some of his staunchest allies, including Peter Cooper, were also badly hit. Field's reputation from his previous financial troubles stood him in good stead, and he arranged matters so that the business could again carry on profitably in his absence. Even more had now come to rest on the 1858 cable. The Atlantic company had spent £301,000 by the end of 1857, and had only £46,000 left in its coffers. To finance the 1858 attempt, new shares were to be issued at £20 each. Finding subscribers, though, after the unexpected failure, was more difficult than before. Nonetheless, an order for 900 miles of new cable was placed with Glass & Elliot. Added to the lengths stored in Plymouth, that made more than 3,000 miles in total, so that the line could be laid generously without a danger of snapping or shortage of cable. Over the previous months, Professor William Thomson had become convinced that the purity of copper in the core was crucial to the success of the cable. The additional lengths made for the 1858 attempt were therefore tested for their quality and conductivity as they were being made, a system rigorously followed ever after. But this happened too late to improve the main line in store at Plymouth.

Gisborne made another attempt to involve himself with the cable. His offer to lay it the following summer, but only on condition that he was entrusted with 'the entire charge of the vessel', was declined. Field, bouncing back from his personal troubles, again visited Washington and was given the services of the *Niagara* and Chief Engineer Everett, who had impressed him in 1857. On his return to England in January 1858 with Everett, Field was appointed general manager of the company, in charge of the entire staff of engineers and electricians. He turned down the offer of a salary of £1,000.

Everett spent the next three months working at the Southwark factory of the engineers Easton & Amos on a re-design of the paying-out machinery. It was considered an immense improvement, smaller and lighter than the previous version, taking up much less space on deck, and a quarter the weight. The machine incorporated a self-regulating brake which could release quickly to prevent the cable from snapping. Everett invited a number of eminent engineers to inspect his machine, including Brunel, Penn and Field, each acting as an unpaid advisor to the Atlantic company. All were impressed by this machine, which 'seemed to have the intelligence of a human being, to know when to hold on, and when to let go'. There had been other improvements to the machines on board: Easton & Amos's engine for coiling on board the *Agamemnon*, Glass & Elliot's hauling-in machine on loan at no charge, and Mr Robinson's new graphite batteries. Most significantly, Thomson had developed new and delicate instruments to revolutionise long-distance signalling and electrical testing on board ship. His mirror galvanometer used a tiny magnet fixed to a mirror, both suspended by a silk thread, to enhance weak incoming signals by light and reflection. A 'mirror clerk' could read up to twenty-five words a minute, dictating the message to a second clerk.

This became the main instrument used in the industry for the next fifteen years, until superseded by Thomson's siphon recorder. The mirror galvanometer's very sensitivity, though, made it unsuitable for use on board ship during laying, when galvanometers were used for electrical testing and as receivers to try the function of the cable as it was laid. One problem was the ship's motion. The instrument also suffered from the magnetic problems which interfered with the working of ships' compasses – the changing magnetic field of the earth, and the influence of the ship's iron hull and cargo. Thomson developed a marine galvanometer which would work under these conditions, by suspending the mechanism to isolate it from the movement of the ship, using magnets to compensate for external influences, or encasing the instrument in an iron shield.

The ships, chief engineers and electricians would be almost the same as on the 1857 expedition, with the *Gorgon* replacing the *Leopard*. At the last minute it was discovered that the *Susquehanna* was stranded in the West Indies, quarantined with yellow fever on board. It was too late to acquire another American vessel, so the intrepid Field immediately took a cab across London to the first Lord of the Admiralty to appeal for help. The Admiralty was short of vessels at the time, but such was Field's audacity and reputation that he managed to secure the Royal Navy's *Valorous*.

The *Niagara* and the *Agamemnon* spent the whole of April and part of May 1858 at Plymouth receiving the cable. This time, it was coiled around large cones, and stowed with greater care than in the previous year. The engineers had their way, the naval officers' advice was to be followed, and the line was laid from the middle of the ocean. One of the advantages in this plan was that the laying vessels could talk to each other, down the very cable as it was laid. Because such a difficult mid-ocean splice had not been tried before, there was to be a rehearsal. The ships set sail for the Bay of Biscay, where the sea was 2,500 fathoms, as deep as anything which would be encountered in the Atlantic. There they made sure that the crew knew their duties, and they practised splicing, laying, hauling in, and buoying, using old and damaged cable. There were breakages, to be expected with the flawed wire, but the results, and the performance of the new instruments and machinery, were pleasing. The ships were able to talk to each other throughout the exercise, with 'perfect electrical continuity'.

The squadron left Plymouth for their mid-ocean rendezvous on 10 June, seven weeks earlier than the previous year's expedition and with much less ceremony. The weather was glorious. Within hours, though, the situation took a dangerous turn for the worse. Maury's careful climatic calculations had been upset, and upset with a vengeance, by a freak storm, a gale so severe that it was called a tempest. The *Niagara* bore the brunt of this violent weather without too much danger. But the ungainly *Agamemnon*, 'labouring fearfully', top heavy with 250 tons of cable on her forward deck, took it much worse and almost turned turtle. The storm lasted a week before reaching its climax, and for two further days the *Agamemnon* was in ever greater peril, in 'as fierce a storm as ever swept over the Atlantic'.

The position of her main cargo in the centre of the ship had shifted as the vessel lurched from side to side, almost on her beam ends, in the towering waves. The ship's planks gaped an inch apart under her huge load, and it was feared that she would break up. The cable, if it slipped, might take out the *Agamemnon*'s side. Her coal bunkers burst, releasing 100 tons of fuel and increasing the danger, the ship at this point lurching 45 degrees with each wave. The cable in the main hold worked loose, 'resembling nothing so much as a cargo of live eels'. If the masts had given way, the ship would have been tossed further to the side and would certainly have been lost, her cable and crew with her. Captain Preedy was left with no choice but to run before the storm, a course which carried the risk that the *Agamemnon* would be swamped, 'pooped by the monster seas in pursuit'. He thought of throwing the cable overboard to save his vessel, but gambled on retaining it. The ship was within hours of destruction when at last the storm blew itself out, and on 21 June she was able to make her way towards the *Niagara*. The cable expedition mustered at the rendezvous point on a still evening four days later.

The *Niagara* had suffered minor damage, and lost the buoys used to suspend the cable. The *Agamemnon* was in much worse shape, hardly seaworthy and with a number of men seriously injured and traumatised. The 'live eels' needed re-coiling, a considerable task which took several days. Once that was dealt with, the expedition could proceed, despite the British ship's condition. The *Niagara*'s cable was taken on board the *Agamemnon*, a splice made, and the ships steamed away towards their home countries. Six miles apart, they were obliged to turn round and start again. The *Niagara*'s cable had been paid out too slack, fouled on its pulley, and broke.

Thomson had taken the place of Whitehouse at the last minute as chief electrician on the voyage. The professor was later thanked by the company for having 'generously given up the whole of his arrangements for the summer', though in truth he was delighted with a chance to carry out electrical experiments impossible in a laboratory. Whitehouse had excused himself because of illness, 'supported by medical certificate' as he was later at pains to emphasise. This was not the full story, for the highly strung Brighton surgeon probably sensed by this point that his star was on the wane. His scientific views were increasingly at odds with those of Thomson, and he had lost his ally on the main board with the removal of Morse as a company director.

On board the *Agamemnon*, Thomson continuously tested the cable as it was laid from the second splice. With eighty miles paid out, he had to announce that there had been complete electrical failure. Before thorough tests were completed, it proved impossible to hold the line from a stationary vessel in a heavy swell, and the cable snapped and was lost. It seemed that the fault had been on the seabed, a conclusion which disheartened everyone involved, for it suggested circumstances beyond control, and which could prove fatal to any attempt.

But there was only one thing to be done. A third start was made on the next day, 28 June. By this time all were exhausted and frustrated, not least because the prolonged voyage meant that the *Agamemnon* had run out of fresh food, and the crew was reduced to eating beef 'salted to an astonishing pitch, and otherwise uneatable, kept three years beyond its warranty for soundness'. The cable was again launched, and the ships parted, soon lost to each other's view in a cold fog. This was almost the last chance, for the *Agamemnon*'s coal was running out as fast as her other supplies. The ships moved slowly, at two knots at first, gradually picking up speed with each hour. Charles Bright and his assistants Samuel Canning and Henry Clifford stood constant watch on the paying-out machinery.

Two days passed, the weather stayed calm, and some of the exhausted men snatched a few hours' rest, lulled by the steady turning of the machine. Mechanically, it was going well. Electrically, the signals passed unhindered. When the line on the upper deck was almost gone, 112 miles from the start, the machinery was slowed so that cable could start to be drawn from the main hold. This procedure had been practised in the Bay of Biscay and should have been quite straightforward. Out of the blue, though, inexplicably, the cable which had just left the ship snapped, while subjected to a force of less than a ton. An explanation came later: the lowest coil of cable had been damaged when the *Agamemnon*'s upper deck floor was smashed in the storm, and the problem passed unnoticed in the chaos around.

There was enough cable left to try again, although coal was very low. The *Agamemnon* sailed, wind-powered to save coal for the laying, for another mid-ocean rendezvous with her American counterpart. The *Niagara*, with no means of hauling in her cable, had been forced to cut and abandon it. On 1 July, the weather worsened, but this was a nuisance rather than a danger as the loss of her upper-deck cable had made the *Agamemnon* 'as buoyant as a lifeboat', removing the dangerous tendency to roll over. As fog descended, she sought the *Niagara* for several days, but failed to find her. Communication between the ships had stopped with the breaking of the line. The Americans, it turned out, had gone to Queenstown, following a pre-arranged order to return to Ireland if the ships had travelled more than 100 miles when communication broke off. The *Niagara* arrived in port on 5 July, where there was great alarm as days passed and the British ship did not appear. But a week later, after thirty-three days at sea and having survived great peril, the *Agamemnon* cast anchor in Queenstown harbour.

Field, who had been on board the *Niagara*, was already in London, confronting a gloomy board of directors. The American, who had been running on nervous energy and little or no sleep, was close to collapse. 'The strain on the man was more than the strain on the cable, and we were in fear that both would break together,' wrote his brother Henry. The directors were in despair, staring blankly at each other. The chairman, Sir William Brown, proposed that the scheme should be abandoned, the cable sold and the proceeds divided among the shareholders. But there was still conviction that the cable could be laid that year and Brown, out-voted, resigned. The expedition was told to prepare to sail again immediately with its remaining 2,200 nautical miles of cable.

Freshly coaled and provisioned, the small fleet turned again towards the Atlantic Ocean on 17 July. There was no cheering or encouragement from the shore. Instead, many were convinced that the idea was hopeless and that the 'stubborn ignorance' of pursuing it was the folly of a company 'possessed by a kind of insanity'. As the ships passed across the ocean, it fell calm, 'smooth as a mill pond'. They rendezvoused on the morning of 29 July. Unceremoniously, for the weary party had no enthusiasm to mark the moment, the cables were joined again. As paying out started, a great whale approached at speed and grazed the *Agamemnon*'s cable where it entered the water. This hazard, unforeseen, caused no harm.

The cable went steadily out. On the British ship, a damaged piece of line was discovered in time to repair just before it disappeared into the ocean. There were interruptions in the electrical current, which righted themselves as mysteriously as they had appeared. Later an inadequate sand battery was blamed for this inconsistency. As laying proceeded, the signal was encouragingly robust. On the *Niagara*, there was alarm as the ship deviated miles off her course, the compass thrown out of true by the mass of iron on board. Gaining distance at that rate, she would not have cable to reach Newfoundland. The problem was solved by sending *Gorgon* ahead to lead the way. The *Agamemnon*, meanwhile, struggled eastwards in high seas and against the wind, narrowly avoiding collision with an American schooner, *Chieftain*, which had ignored warnings to move out of her way.

As the American cable-layers passed their first icebergs, they received a signal that *Agamemnon* had paid out 780 miles of cable. Both ships had then, on 3 August, reached shallow water of 200 fathoms. The following day, *Niagara* entered Trinity Bay, and the pilot steered her to the Telegraph House. Cyrus Field telegraphed the news of her arrival to the Associated Press in New York early on 5 August, just before the cable was landed. At the same time, *Agamemnon* was entering Valentia harbour, heralded by the guns of the *Valorous*, which roused the sleeping inhabitants. The boy John Lecky was among those who rushed out to witness the *Agamemnon*'s return to Valentia, one year to the day after she had last been there, a year of 'profound calm' on the island. He knew, though, that recent weeks had been anything but calm for the naval vessel, and that she had been almost lost with her cable. Now, however, she was 'lying like a cork' in Lough Kay, 'having laid her part of the cable from mid-Atlantic and having burnt all the coal as well as some of her decks, she was wonderfully light and it is a wonder how she got in without disaster.' Bright and Canning took the cable ashore at Knightstown, where

Whitehouse attached a galvanometer and sent the first message between the continents.

News of the triumph had been telegraphed to every part of North America after Field's message. 'The impression of this simple announcement it is impossible to conceive. In some places all business was suspended; men rushed into the streets, and flocked to the offices where the news was received.' At a religious gathering in Massachusetts, 1,000 people rose to their feet and cheered at the news, then gave thanks to God for this great feat, 'calculated to hasten the triumphs of civilisation and Christianity'. A hundred guns were fired on Boston Common, and New York, once it could believe it, broke into 'tumultuous rejoicing'. The newspapers talked of nothing else, and Cyrus Field, as he later put it, 'awoke and found myself famous'. A modest man, he was careful to credit the work of his 'co-workers' in the achievement. But Field was the figurehead, and within hours 'his name was on millions of tongues'. Congratulatory telegrams flooded along the New York and Newfoundland line to Field, including ones from President Buchanan, from Peter Cooper and from the Governor General of Newfoundland. Some came from unlikely quarters: 'Beech Tree, chief of Oneida tribe, honors the white man whom the great spirit appoints to transmit his lightnings through the deep waters.' The telegraph clerk, plainly unfamiliar with a third-person form of address, had added at the bottom: 'No signature.'

Charles Bright communicated with the Atlantic company board as soon as the *Agamemnon* reached Valentia, reporting good signals with the *Niagara* in Trinity Bay. The engineer and his colleagues then enjoyed their first sound sleep for weeks, before joining in a round of civic banquets and celebrations in Valentia, Killarney and Dublin, where the Lord Lieutenant of Ireland, on behalf of the queen, knighted the twenty-six-year-old engineer. Bright became the youngest man for generations, as well as the first electrical engineer, to receive such a distinction.

No messages were allowed to pass on the cable until the electricians had finished their tests. The public was impatient, even though most would never be able to afford to use an Atlantic telegraph personally. Yet they longed for news, for confirmation that the miracle really had come to pass. The silence was broken on 16 August, when the directors cabled their American associates: 'Europe and America are united by telegraphic communication. Glory to God in the highest, on earth peace, goodwill towards men.' The next message was a congratulatory note of 98 words from Queen Victoria to President Buchanan, which in the checking and rechecking took sixteen hours to transmit. Once the White House had convinced itself this was not a hoax, the president penned a reply of 149 words, sent in ten hours.

Confirmation that the cable worked was a cue for further outpourings of joy. In cities across the United States, guns were fired, flags flew and church bells rang. New York was illuminated on a scale never before seen, 'as if it were intended to light up the very heavens'. This brought further drama when the blazing lights of City Hall set fire to its cupola, which was destroyed. The hall itself narrowly escaped being burned to the ground.

Cyrus Field had arrived home with the *Niagara* on 18 August, having just resigned as general manager of the Atlantic company, and hoping for some rest and peace. He was thirty-eight years old, his sixth child had been born the previous year, and he had barely seen his family for years. First he had to accept a round of celebrations and ceremonies in keeping with his new heroic status. But once this had died down, there was still to be little repose, for there was something not right with the cable. Field had been concerned from the cable's first days that messages could not be transmitted fast enough. As the trials continued, the speed did not improve, and he feared that the telegraph would not be commercially viable. This was disquieting, but it was swiftly overtaken by a greater worry still.

Charles de Sauty, the electrician left in charge of the Newfoundland cable station, was having increasing difficulty transmitting and receiving signals. Sometimes the line worked well enough for the operators to chat with each other. One or two highly significant messages passed, for instance news that the Sepoy Mutiny in India had been put down, which halted the mobilisation of two British regiments from Canada. Nine words on the cable had saved the British government perhaps £60,000. In twenty days, the cable carried 271 messages, at an average of ten words each, from Newfoundland to Valentia; in the other direction passed 129 telegrams. On some days, though, de Sauty could receive nothing. The problem seemed to lie at the Valentia end of the cable, where Whitehouse was busily conducting experiments of his own.

John Lecky's father had cleared machinery out of a disused slate-sawing yard and lent his building there to the telegraph company. With instruments installed in the east end of the shed, overlooking the pier, this served as a telegraph office for as long as the cable worked. Thomson's delicate reflecting galvanometers were available, but Whitehouse had chosen to use his own 'detector' instruments. These needed a vast increase in the electrical charge:

> Unfortunately for the life of the cable, Mr Whitehouse was imbued with a belief that currents of very high intensity, or potential, were the best for signalling; and he had enormous induction coils, *five feet long*, excited by a series of very large cells, yielding electricity estimated at about 2,000 volts potential. The insulation was unable to bear the strain, and thus the signals began to gradually fail.

Blame for the damage landed squarely upon Whitehouse, perhaps unjustly, as many of his views had been supported by his old ally Morse. The cable was far from perfect when handed over to Whitehouse, despite everything the public, and most of the directors, had believed. But Whitehouse's actions, described as 'high pressure steam got up in a low pressure boiler', further damaged the insulation and hastened the telegraph's end. Most expert opinion of the time concurred in this view. Whitehouse was obliged to revert to using Thomson's equipment, powered by Daniell cells, and then the professor himself arrived and tried to coax the cable back to life. A few more messages passed, but effectively it was all over, even before the climax of the New York celebrations.

The directors embarked for Valentia for the formal opening of the line to public messages. Instead they found themselves alongside scientists trying to work out what could be done to save the cable. 'Everything possible was thought of and tried but to no purpose,' noted John Lecky. 'They tried a very powerful current and my father had a huge slate battery made for them. I remember one day going into the operating room and one of the electricians took up a barrel steel pen by an insulated tongs and sent the current through, when the steel melted like wax in a candle!' After a time the office, with some of its equipment, was abandoned to the boy.

The fatal defect appeared to be less than 300 miles from Ireland, but there were not then the means to grapple and repair the cable. Even if it could have been retrieved and mended, the line would most likely have quickly failed elsewhere. Its faults were deep-rooted, stemming from the first stages of design and construction, and compounded by its handling afterwards. The cable was by far the longest line ever made, and though its makers were the best in their field, all were novices. The firms of Newall and Glass & Elliot operated quite separately, not communicating with each other as the line was manufactured, and certainly not pooling experience or ideas. As failure cost them nothing, they had no immediate incentive to develop better procedures. The cable made in 1857 was not tested during manufacture – it had been left to Professor Thomson to identify this as a failing and insist on new procedures. Perhaps most damaging, though, was the treatment of the line in the year between its manufacture and laying, when it 'suffered enough to have destroyed its efficiency even had it been free from faults at first'. The gutta percha insulation had deteriorated when stored outdoors in the sun. The cable was then coiled and re-coiled as it was loaded on board ship and taken off in 1857, then stowed again in 1858 and further damaged in the storm. 'Take all these things together, and the wonder is, not that the cable failed after a month, but that it ever worked at all.'

Laying the cable had been an unprecedented feat of engineering, and the telegraph proved its worth by carrying messages of high importance during its short life. Charles Bright's triumph, against the odds and contrary to a great weight of expert opinion, was a tribute to the young man's ability and determination. Yet the cable had failed despite this, and no one could agree about the causes. The line was summarised at the time as 'mechanically good but electrically bad'. The success of Bright and his fellows had briefly diverted attention from underlying electrical questions, and disguised just how unsophisticated and experimental electrical understanding and technology were. Until these questions were resolved, the line would never work.

5

Languishing

by Delay

The Atlantic company reached the end of 1858 having used up all its assets, and with more liabilities than it could meet. Even if the cable could have been repaired, there was no money to attempt it. All staff were under notice of dismissal. A persistent myth that the 1858 cable was a fraud and had never actually worked at all was widely believed, despite evidence to the contrary. As one commentator put it at the time, 'When engineers have no definite plans, directors no capital, and the public little or no hope, joint stock enterprises have not a brilliant time of it.'

A total of £465,000 had been spent, most of it a total loss, on the Atlantic dream. Before turning to any positive lessons which might have emerged, it was time for recriminations. Whitehouse was set up to take a great portion of the blame. Thomson tried to defend him, for he thought the man industrious and well-intentioned, but the more the professor investigated what had passed, the more the other's inadequacies were exposed. The directors accused Whitehouse of ignoring their instructions and wasting thousands of pounds on his own experiments. He was immediately dismissed as electrician, but would not go quietly. Attempting to defend his personal and scientific reputation, he had by the end of September published a booklet setting out his case. He railed against 'sinister forces' which had unseated him, particularly 'the frantic fooleries of the Americans in the person of Mr Cyrus Field'. And this was not the last, for Whitehouse appeared at an extraordinary general meeting of Atlantic shareholders in December 1858, to suggest a merger with a new company. This new venture, he said, had already raised half the capital needed for another cable. The chairman, James Stuart Wortley, politely silenced Whitehouse. Later, in private, Whitehouse declined to give details of this mysterious rival scheme to Stuart Wortley.

This may have been no more than an attempt to strike back at his persecutors, a figment of Whitehouse's imagination. But there certainly were other projects competing with the Atlantic company, and some of them were making serious progress. Cyrus Field claimed to be unconcerned by these various schemes, arguing that there was enough business for more than one line. The danger to the Atlantic company, though, was that another enterprise would swallow any funds available, and that Field's project would be starved of investors.

The holy grail which all pursued was a route to avoid the worst physical dangers of icebergs and volcanic activity in the Atlantic, and preferably one where relays could be employed to break the distance. There had been such a scheme floated between the two Atlantic attempts of the 1850s, in the spring of 1858, by the European & American Submarine Telegraph Co. The promoters, mainly French, claiming concessions in France, Spain and Portugal, tried to raise £1 million for a telegraph from Plymouth to Finisterre, Lisbon and the Azores, then onwards either to Boston, or to North Carolina via Bermuda. For as long as the fate of the Atlantic company's cable was uncertain, investors were reluctant to support this, but the plan re-surfaced later.

After the failure of the Atlantic cable in 1858, others too were inspired to try. The North Atlantic Telegraph Co. used the events of 1858 to argue that 'a direct line could not be expected to succeed as a commercial telegraph'. On their behalf, Colonel Tal P. Shaffner, an American adventurer encouraged by the Danish government, chartered a boat to survey a route from the north of Scotland by way of the Faeroes, Iceland and Greenland to Labrador or Newfoundland. Although the inhospitable climate would have made it difficult to lay and maintain cables, not to mention recruit staff for the telegraph stations, the scheme gathered some momentum. Sir Charles Bright was appointed engineer. The Prime Minister, Lord Palmerston, arranged Admiralty help for Shaffner to make ocean soundings in 1860. The paddle steamer *Bulldog*, under the experienced Arctic navigator Captain Leopold McClintock, was despatched to work for more than four months alongside the company's own boat, the *Fox*. The results were encouraging, and Bright and McClintock presented their findings to the Royal Geographical Society in January 1861. The project faltered, though, for finance, a consequence of the collapse of confidence by investors after 1858.

The alternative was to go by land, the long way. An idea came together in 1862 for a trans-Siberian line to link

New York with London by crossing three continents. Perry Collins, the United States commercial agent at the mouth of the Amoor River in eastern Siberia, had been trying to promote a cable from there to San Francisco. The Tsar, meanwhile, planned to extend the Russian telegraph network eastwards towards Amoor. Collins managed to interest the American Secretary of State in a larger scheme, offering benefits for trade between the two countries, as well as a valuable link between America and Europe. A preliminary survey brought home the size of the task. As there was still no telegraph east of Moscow, at least 14,000 miles of new cable, most of it overland, would be needed. California had been reached by cable from the east in 1861. While San Francisco is 5,000 miles from Amoor, the Bering Strait measures a mere forty miles, and this would have been the only submarine section needed on a line connecting New York with Paris. Hiram Sibley of the Western Union Telegraph was sure that he could complete the project in two years, and perhaps in one, and Samuel Morse predicted no major difficulties. The great advantage of this scheme was its cheapness, only $1.5 million.

The Russian government was supportive, and the promoters played down the magnitude of the operation. While the capital costs were relatively low, running expenses would be vast, as they needed to support dozens, maybe hundreds, of telegraph stations. But the failure of the Atlantic company, far from undermining this scheme, seemed to give it greater credibility:

> This question has been reduced to a far less gigantic task than it seemed only a short year ago. The only subject of

discussion now being the least length of deep sea cable. We have already seen gigantic efforts made, and immense sums of money cast, as it were, into the ocean in futile attempts to connect great distances. The Atlantic cable, unhappily, after exciting the enthusiasm of the world, has come to a disastrous end; the great Red Sea and India telegraph has also succumbed to a fate quite as lamentable. We name these facts, still not without hope, that science, ingenuity and indomitable will may yet overcome all obstacles. The recent success of the Malta and Alexandria cable has inspired the friends of long submerged lines with new ardour and ambition for further efforts.

But, readers were reminded, the maximum length of any section on that last line was a mere 300 miles. Long submarine cables were still seen as a very bad risk.

In spite of all these grand plans, three years after the Atlantic failure there was little real progress on any. George Saward, secretary of the Atlantic Telegraph Co., summed up his feelings as the tenth anniversary passed of Brett's cross-channel cable:

The history of submarine telegraphy – fairly written – would reflect little credit, either scientifically or morally, upon the present age. In truth, the business of making and laying telegraphic cables is – after a lapse of ten years since their introduction – only now commencing to be treated with the dignity of a science. Great and useful principles have been elicited by repeated failure, but they have been mostly ignored in practice.

There were two particular groups who angered Saward. One was those scientists who cloaked their discoveries in mystery, men too wrapped up in 'speculative philosophy' who hid 'the grain of scientific wheat in algebraic chaff and mystification'. In fact, thought Saward, 'the successful construction of the telegraph is no mystery to any mind of fair capacity, depending as it does simply upon the practical application of a very few easy and perfectly well-known natural laws'. In this, the company secretary tended to oversimplify. But nor did Saward have patience with the commercial men who had brought matters to a deadlock because they had failed to reflect sufficiently on the technical aspects of the business. He had particular scorn for the way 'projectors, concessioners and dealers in monopolies have been allowed to mix themselves up as directors, contractors or officers in these undertakings'. Saward was furious and frustrated, because everyone who knew anything about the subject knew by then that an Atlantic cable could and should work, yet no one seemed to have the wherewithal to make it happen.

Field had had other things on his mind. He was forced to stay in New York to concentrate on his own business after a costly fire early in 1860 for which he was not fully insured. It was during this period that he first saw Brunel's *Great Eastern* afloat. The massive ship was in America in an attempt to cover some of her losses by carrying fare-paying passengers, and Field joined a short cruise. As a holiday, it was a disaster, for the vessel was not organised nor equipped for its numbers of passengers, but the experience underlined the *Great Eastern*'s potential as a cable-layer. Field's financial position worsened as 1860 progressed, and he was obliged to mortgage all his personal property to save the paper business.

He still had a lucrative share in the American Telegraph Co., which allied itself with Morse's own lines to secure a monopoly in the eastern United States. This new operation joined with Associated Press to run a telegraph and fast steamer service from Newfoundland, based on Gisborne's original relay plan, which cut two days from the previous fastest time for transatlantic messages. It was no substitute for an ocean cable, but it helped whet the public's appetite for ever faster news services, a clamour which intensified after the outbreak of war in the United States.

Though business matters had stalled, there had already been much headway on the scientific front. At first after the failure of 1858 there had seemed to be more questions than answers. The Atlantic company had hopes of repairing the line, which was believed unbroken, although controversy continued about the causes of its demise. In 1860, the directors sent an expedition to Newfoundland to recover a length of the cable there. Only five miles was fished up, and although the seabed was found to be much rougher and more uneven than expected, the cable itself still worked well. The puzzle over its failure seemed no nearer to solution.

Yet the cable, as Field said later, 'did more by his death than his life'. In its failure, he argued, it had been of immense benefit to science. 'It has been the great experimenting cable. No electrician ever had so long a line to work upon before, and hence the science of submarine telegraphy never made such rapid progress as after that great experiment.' Every cable subsequently laid, where the projectors took lessons from the 1858 line, had been successful. 'All these triumphs over the sea are greatly indebted to the bold attempt to cross the Atlantic [in 1858].' This, though, was written with long hindsight.

At the time, over the winter of 1858 and through 1859, Field struggled to keep Atlantic cable hopes alive. He described these as his darkest days. 'It is more difficult to revive an old enterprise than to start a new one.'

Recognising that private investors had been frightened away by the scale of losses, Field saw a way forward through attracting government support. But this was not easy in the circumstances of 1859. All governments expressed interest, sometimes even enthusiasm, for international cables, but none was prepared to risk the sums needed to finance such schemes. Britain was particularly keen to improve communications, especially in the direction of her empire in the east. She was also markedly reluctant to commit public money to the enterprise. The best that companies could hope for from government was guarantees to their shareholders, and these were conditional on the line being successfully laid. It was still thought that long submarine cables may not prove profitable, so that dividends needed to be assured. But a guarantee on those terms did nothing to address the main problem, that investors lost everything if the cable itself did not work. And after the Red Sea debacle, the conditions of government support tightened further.

The efforts to revive the Atlantic scheme after 1858 took place against the backdrop of another submarine disaster, that of the Red Sea cable. The Red Sea telegraph had promised a secure route from Britain to India, avoiding the practical and political problems of a line across the Ottoman Empire. The overland telegraph was unsatisfactory even in peacetime, with Morse code messages sent on by telegraphers who did not understand English. Telegrams were delivered days or weeks late and so full of errors that they were often impossible to decipher. Events of 1857 underlined the urgent need for new arrangements. During the Indian Mutiny, an emergency request for troops had taken forty days to reach London from Lucknow. Lord Derby's government therefore decided to back a Red Sea cable, which would have the advantage of being under British management and control, so that foreign powers could not intercept confidential messages nor interrupt the flow of information.

Although the government was keen to have this line in place, they would commit themselves only to a guarantee to shareholders of a minimum four and a half per cent return. The condition was that the telegraph should test well for a month after laying. In theory, if the line failed, there would be no call on public funds. If it worked, it should make at least some profit and the worst case would be that the British government subsidised a strategically important cable. This agreement appeared as watertight as the cable, binding the government to pay only if the cable were successful. Unfortunately neither worked at all as intended.

The submarine line, in six parts totalling 3,500 miles connecting Egypt with the west coast of India, was completed by Newall in February 1860. Some of the sections worked for more than two years, but the entire line, from Suez to Karachi, operated for only a few hours. No telegram ever travelled the whole distance, and it soon became apparent that there were serious problems with all six sections. The line had not been tested underwater before laying. Its protective iron armour was too light and soon rusted, allowing worms to eat through the insulation. Furthermore it was laid too tight on an uneven seabed, so that when it became encrusted with barnacles the extra weight caused hanging sections to break. There were no arrangements in place for repair. All in all, the enterprise was described as 'like running a donkey for the Leger'. The promoters had fatally underestimated the scale of their task.

Yet the shareholders did not lose. Every section of cable had tested successfully, as required in the contract. Derby's successor, Palmerston, and his chancellor, Gladstone, were therefore bound to pay the maximum penalty, £36,000 a year, despite having no working cable to show for it. Over the following fifty years, the cost to the Exchequer was £1.8 million.

Stuart Wortley, the Atlantic company chairman, had already, early in 1859, secured promises of subsidies from the Treasury. There was an offer of up to £20,000 a year to cover the cost of official messages on the line, and a guaranteed return to shareholders of eight per cent for twenty-five years on a new capital of £600,000. In exchange, the government wanted the exclusive Newfoundland concession, originally negotiated by the New York and Newfoundland company. But none of this would come into effect until the cable was working. So much money had been lost to date, and public confidence was so low, that a conditional subsidy like this could not tempt investors. But the government, after its embarrassing loss in the Red Sea, would go no further.

In the years following 1858, while little progress was made with any transatlantic project, the cause of submarine telegraphy advanced steadily in the Mediterranean and Middle East. The Atlantic attempts attracted most publicity, and unnerved investors. The eastern expeditions, though, with less fanfare, were crucial in developing cable technology during the early 1860s. Much of the Mediterranean work was contracted to R.S. Newall. His Malta-Cagliari-Corfu line gradually failed between 1857 and 1861, and the Malta-Alexandria telegraph frequently broke down, yet the cables themselves were improving significantly, as were laying and testing techniques. The Malta-Alexandria telegraph of 1861 was in fact a landmark, the first deep-sea cable successfully laid. It was the first sent out in tanks and tested under water on board ship. Although laid in water that was too shallow, so that it needed constant repair, it worked well for ten years. The lessons learned in the Atlantic were finally being put into practice, yet the public remained wary of submarine telegraph shares. The engineers and large investors though, closer to events, were convinced that the technology was now viable, and long-distance lines would be highly profitable.

Field's 'great experimenting cable' had given Professor Thomson and his colleagues much to think about. Thomson and Fleeming Jenkin, previously a telegraph engineer for Newall, started to design sending keys which would answer the problem of sending messages down long cables at commercial speeds. Eventually they came up with an automatic sending device. Thomson and Jenkin established a partnership with Cromwell Fleetwood Varley, inventor of the 'curbing capacitor'. The electricians discovered that by using these instruments together, and introducing duplexing, or two-way traffic on the line, the problem of 'retardation of the signals' was effectively solved. Speeds greatly increased after 1860, up to perhaps twenty-five words a minute, though before 1873 between seven and thirteen was a more usual rate.

While there was a hiatus in operations in the Atlantic, the British government and Atlantic Telegraph Co. were not inactive. The government took a course which turned out to serve the industry better than any number of financial guarantees could have done, one which effectively settled the main scientific controversies surrounding long-distance cables. An enquiry was set up under a respected Board of Trade technical expert, Captain Douglas Galton, a combined investigation by the Board of Trade and the Atlantic Telegraph Co. Galton's deliberations were seen as 'the most valuable collection of facts, warnings, and evidence ever compiled concerning submarine cables'. Galton himself believed that if cables were reliable, government money would not be needed in any form, for private investors would take up the job of laying submarine lines. His committee met between December 1859 and September 1860, and in 1861 published a report with recommendations on the making, laying and working of undersea cables.

During the course of his enquiry, Galton interviewed most of the leading submarine telegraph engineers of the time. He collected evidence about the abortive attempts on the Atlantic, and about the more and less successful cable-laying experiences in the Mediterranean and elsewhere. Thomson, Jenkin and Varley were key to his investigation. All were respected scientists with recent practical experience of cable-laying, and understood mechanical as well as electrical engineering. Jenkin's experiments in Newall's factory during 1859 had proved pure gutta percha to be a generally better insulator than any other compound or mixture. The young engineer could also suggest ways of avoiding faults in submarine cables by improvements in manufacture and laying, and how to mend them when they arose. The versatile Jenkin had also been investigating Thomson's theory, the 'Law of the Squares'. He presented Galton with proof of the law, showing that increased input voltage had no effect on the rate at which messages could be transmitted. Whitehouse still refused to accept the 'Law of the Squares' and continued to argue that higher applied voltages would force signals along

Atlantic cables. The company electrician's decision to apply hundreds, if not thousands, of volts to the 1858 cable had had catastrophic results, but he still blamed the poor standard of the cable itself. Thomson and Jenkin finally and indisputably proved the folly of Whitehouse's actions.

Galton's final report listed the submarine lines laid between 1850 and 1859. At first reading, it is a dismal picture. Of the many miles laid down, a great proportion had failed. The unsuccessful ones were mainly deep-sea cables that had had mechanical failures, usually a breakage. Galton thought that the early successes of submarine telegraphy, the cables between Dover and Calais, and Orford Ness and The Hague and Ostend, had given false confidence, for they actually concealed serious gaps in understanding and expertise. He concluded that there must be much more careful quality control during cable construction, and also much greater care in cable-laying. This was not entirely new to the industry, but clearly the best practices of leading engineers needed to be adopted universally. Galton's other recommendations were for standardisation, in electrical units and measurements, and in signalling and receiving. Much of this came to pass during the early 1860s under the guidance of Thomson, Jenkin and Varley. The vital but unglamorous standardisation of the ohm and other units continued quietly through the early 1860s, guided by Jenkin and the British Association for the Advancement of Science. Galton had exonerated Thomson and his circle, and because his investigation had been so searching, it settled the matter in their favour. Galton was decisive. In the dispute between Thomson and Whitehouse, there could no longer be any doubt about who was in the right.

Galton continued to be a key government advisor on the subject of deep-sea cables. The Atlantic Telegraph Co. was still in the financial doldrums, and still seeking public help. In March 1863 another approach was made to Gladstone. The company had reduced the price of new shares to £5 in the hope of appealing to a wider group of shareholders. Prospective investors were told that the reasons for failure in 1858 had been 'fully ascertained'. The faults:

> arose from no inherent scientific difficulties, *even then*, when knowledge on the subject was so far in the rear of its present position. The result was simply what might have been expected, or, to speak more correctly, it was even less disastrous than might have been reasonably looked for from imperfect manufacture – the want of due superintendence and good organisation, and the absence, owing to the then state of science, of the means of testing the cable with anything like the accuracy that is now possible. These elements of failure no longer exist, and in particular the various improved methods at present in use for testing submarine telegraphs are such as readily to detect imperfections which, owing to their minuteness, must formerly have escaped observations altogether.

The company was confident in the technology and expected success; they needed £600,000, but only £220,000 had been raised. Could the government subscribe £100,000? 'There is danger of the enterprise languishing by delay.' The company's new eight per cent preference shares were guaranteed for twenty-five years by the British government, but only while the cable worked. There were also subsidies each year of £20,000 from Britain, and $70,000 from the US Congress, covering the cost of official messages. Reuter guaranteed business worth £5,000 a year for the line from his news agency. Although the Newfoundland and other landing rights in British North America had been conceded to the British government, the company retained its concession in Maine, and had reached an agreement to link in with the overland telegraph system in the United States. The business was beginning to take shape.

Gladstone referred the company's appeal for funds to Galton, who had moved to the War Office, who consulted Wheatstone and others to assess the risk. They had learned much from the Red Sea failure. Continuing experience had shown the best ways of constructing a cable – the preferred insulation, how best to apply it. The qualities of copper wire were better understood, and so was the need to constantly test cable at every stage of its manufacture. The instruments for working lines had been 'materially improved, and rendered far more sensitive', which prevented damage to the cable. Despite all these positive advances, though, Galton and his associates urged caution, for they saw a weak link in the chain:

> Notwithstanding all the experience which has been acquired in laying cables, considerable risk attends this part of the enterprise. This risk partly depends on the weather, and partly on the degree of vigilance and prudence exercised, at every moment of the day and night, by the persons who superintend the operation. It is a risk therefore which no care and foresight can entirely remove.

His advice went further: 'The risk is much diminished by the use of powerful steamships, properly constructed, and by the persons who undertake the operation holding a large interest in the success of the enterprise, and it must be recollected that with less experience than is now possessed, a cable was successfully laid across the Atlantic'. Galton's counsel was to prove wise.

So the Atlantic company had 'prepared the way for success, by paying for the greatest experiment in telegraphs which the world has known, and now they come forward with their dearly bought experience to renew the attempt to unite Europe and America. The enterprise is of great importance to commerce, and would largely promote imperial interests. It offers strong possibilities of success.' Yet by June 1863, it had raised only £300,000, half of what was needed. The Treasury finally announced that it could not help, telling the company that the issue should be left to 'private capitalists'.

As an incentive to attract more investors, it was decided that the price of telegrams could be doubled to five shillings a word. The existing tariff had been set under an agreement signed before the failure in 1858, when predictions of signalling speed were over-optimistic and before it was appreciated just how few messages might be carried. Under the terms of the financial guarantee, though, the company could not increase the rates without British government agreement. In July 1863 Field wrote to Gladstone, urging a speedy decision on the price increase, so that the new arrangements could be quickly advertised. 'It is of great importance that our efforts should take place before the annual exodus of the wealthy to the seaside and the continent.' Gladstone's officials, however, were privately advising him that the 1858 agreement had lapsed, and there was uncertainty about whether the 1859 contract was still in force. It also came to light that the Atlantic company's rights in Newfoundland had run out after the failure in 1858. Only with difficulty was the landing monopoly there extended to March 1868, on condition that a cable was then in place and working. Although the directors did intend laying a new cable in 1864, this new deadline added to the urgency.

While these developments in Britain were a matter of frustration, across the Atlantic political events threatened to sink any immediate prospect of a cable. The civil war which raged for four years from 1861 in the United States brought personal tragedy to some of the Atlantic company's stalwart supporters. Lieutenant Berryman, a native of Virginia, torn in his loyalties but holding firm to his duty to the Union, died of illness on the battlefield early in the war. His colleague Maury, also Virginian by birth, survived the war, but it ended his naval career. After decades as an officer of the US Navy, he tendered his resignation three days after his state seceded from the Union in April 1861. Appointed a commander in the Confederate Navy, Maury was soon afterwards sent to England as a spokesman for the Confederacy. It was years before he would return to his native land. The conflict also hit Field's American Telegraph Co., which lost much of its line, to the advantage of its rival, the Western Union. Yet the war also gave telegraphs a chance to prove their worth.

As far as Cyrus Field was concerned, the war which could have stopped or delayed the Atlantic scheme, in fact stiffened his resolve. He was diverted for a time into applying his knowledge of telegraph systems to help the Union with its overland communications. After that, he turned to protecting relations between Britain and the United States, anxious to ensure that the British were not drawn in on the side of the rebels. During the early 1860s Field struck up a friendship with John Bright, the Liberal MP, a Quaker who supported the Union because of his opposition to slavery, and who was widely admired in the northern states. Field also became close to certain other American supporters of the Union in London, notably the banking partners George Peabody and Junius S. Morgan. Field at this time too cultivated a regular and friendly correspondence with Gladstone, who favoured the Confederacy.

In 1862, the British chancellor declared himself deeply impressed with religious publications and a book about the misery caused by the war, which Field had sent him. The American acted as intermediary between his government and Gladstone, obtaining copies of the 1862 diplomatic correspondence of the State Department for his friend. In April 1864 Field sent William M. Evarts, a New York lawyer, to brief Gladstone about recent events in the United States. and especially to convey the anger felt at Britain's role in refitting vessels as warships for the Confederacy. Field's excursion into diplomacy was not without cost, for himself and for the cable. At one point he was the target of smears by Confederate supporters in Britain, including some English newspapers, which accused him of using American telegraph companies to line his own pockets. His reputation for probity enabled him to weather that particular storm.

The friendship with Gladstone served no direct purpose for the Atlantic telegraph, but the two men remained close. Their correspondence and meetings endured long after the cable was complete, and after Gladstone became Prime Minister in 1868. Field never gave up his campaign to smooth over misunderstandings between the countries. 'You may rely upon my doing all I can to promote good feeling between England and the US,' he wrote in 1872. Twenty years after the cable was finally laid, Gladstone and Field were still in earnest debate, about the meaning of the first chapter of Genesis.

Field continued to visit London throughout the war, combining his political work with lobbying for the cable. In March 1862, he attended a 'telegraphic soirée' at the home of Samuel Gurney, the Quaker banker, MP and director of the Atlantic Telegraph Co., attended by aristocrats, politicians, financiers, engineers and many other influential members of the pacifist Society of Friends. Wires from four different land and submarine telegraph companies were extended to Gurney's house near Hyde Park, and messages recorded in Morse code on continuous strips. 'Here, for the first time, a gentleman's library was brought into instantaneous communication with all the

The Morse ink writer, 1867. (*Illustrated London News*)

capitals of Europe, Malta, Alexandria and the East.' Stuart
Wortley proved not merely that the 1858 Atlantic cable had
worked – dated publications carrying news from the other
side of the ocean gave easy verification – but also its immense
financial and political value. A renewal of the scheme would
be 'an incalculable advantage to this country', he told the
assembly, but more than that, would serve the interests of
'humanity itself'. Field, for once, said very little, passing over to
the electrician Cromwell Fleetwood Varley to describe recent
improvements in cable-laying and management. Varley, who
had a tendency to overstate the case, believed that the cable
during its brief life had proved so useful that 'if an Atlantic
cable should only last twelve months, it would be cheap to
the country to lay one annually', especially as new techniques
were rapidly increasing the speed of transmission, and hence
the number of messages which could pass.

While the American Civil War added to Field's difficulties
in many ways, it also supplied him with ammunition for his
campaign on behalf of the cable. War brought a stark reminder
of the dangers of misunderstandings between the United
States and Europe. In fact, claimed Field when addressing the
American Geographical & Statistical Society in May 1862,
the war had almost delivered the cable for him. During the
blockade of southern ports by the north, in the early months
of the Civil War, two Confederate commissioners on their
way to Europe had been forcibly taken off the *Trent*, a British
Royal Mail steamer, in the West Indies. When news of this
reached London in November 1861, such was the fury with
the United States and sympathy with the Confederacy that
Britain almost gave up her neutrality and entered the conflict in support of the south. War was averted only by intense

diplomatic efforts. A discreet approach was at this point made to Glass & Elliot by a British government representative, asking how quickly a cable could be made and laid, and at what price. The company asked £675,000 and undertook to complete the line by July 1862. 'Well might England afford to pay the whole cost of such a work', argued Field, 'for in sixty days she expended more money in preparation for war with this country than the whole cost of manufacturing and laying several good cables between Newfoundland and Ireland.'

The immediate danger following the *Trent* incident subsided, and any prospect of financial help for the cable from the British government again faded. But Field saw that he could capitalise on recent events to advance the cable. As the Union began to gain the upper hand in the conflict, Field brought out potent commercial arguments to enlist the support of merchants from the eastern cities. While Britain had most to gain politically from the cable, the United States, he suggested, would enjoy great commercial benefits:

> The shipment of gold, which is constantly taking place, would be much diminished; the rapid fluctuations in exchange would be prevented; and the enormous depreciation of public securities would be much abated. Those speculative transactions in cotton and produce, which have often brought about financial crises in England and the United States, would be rendered almost impossible; and the gain to owners of shipping, on both sides of the Atlantic, would be incalculable, from being able to communicate constantly with their captains and agents in all the ports of Europe and America.

When Field addressed the New York Chamber of Commerce in March 1863 he pointed to the newly opened telegraph to San Francisco, which had 'as much business as it can do, and has earned more than enough to pay the whole cost', even though it connected with a single state of only a few hundred thousand people. As for the Atlantic cable, he did not believe that ten cables could begin to satisfy the demand. 'The great commerce of our ports demands prompt communication with Europe. You cannot write to England and receive a reply under twenty days.' On a single Atlantic line, his conservative estimate was an income of £413,000 a year. This would make possible a forty per cent dividend, though Field had a more ambitious plan. By distributing only eighteen per cent of the gain to shareholders, the rest could be invested in further cables and there would then be nine cables working by 1870, 'without increasing the capital stock at all'.

He was well received, and the meeting passed a resolution stating their belief that the cable should and would be laid, and that the public be encouraged to support it. Subscriptions had then reached £195,000, including £2,000 each from Cooper and Hunt, and a thousand each from a number of other New Yorkers, with a few new supporters signed up after Field's plea. These were useful sums, but nothing near the amounts poured in by the original directors almost ten years earlier.

Field had even less success in Philadelphia and Boston, where he addressed meetings of merchants in April and May 1863. He had hopes that the Boston men, who had contributed largely to railroad schemes, would support the telegraph for the same reasons. He argued that the telegraph would bring American commerce into line with that of Europe:

On the other side of the ocean, the work of extending telegraphic communication has been rapid. I can take my stand in the office of the Atlantic Telegraph Company at Valentia on the west coast of Ireland and communicate with every capital in Europe – to Constantinople, 3,100 miles, and even along the shores of Africa – and as far as Omsk in Siberia, 5,300 miles. Consider the effect upon commerce, upon international trade. Here is Boston, a great commercial city – what an advantage to you if you could instantly communicate with those ports; if you could know when your ships arrive; and the prices ruling in the markets of Europe, so that you could direct from here what cargoes to take on board. Already your great China merchants find the immense advantages of the overland telegraph to the Pacific.

Since the transcontinental telegraph had opened, Boston ships, after sailing round Cape Horn on their way to the Far East, put in at San Francisco to collect the latest orders from home on where to go and what to buy. Finally, Field's guiding principle, that the transatlantic cable would increase international understanding, also presented advantages to trade. He could cite the *Trent* incident, not eighteen months since, when 'England nearly went to war with America because there was not a telegraph across the Atlantic'. But this also presented Field with a problem, for relations between the Union and Britain were not fully repaired, and this stood in the way of American support for the cable. 'One feeling may embarrass the effort to raise money for this enterprise at the present time. It is the feeling towards England on account of her course towards the US since the commencement of the present war.'

Field's audience – men who had been disappointed before by promises that an Atlantic cable was a technological certainty – needed reassurances about the safety of their investment:

> Is such a telegraph possible? It is only a very few years since it was thought possible to lay a cable of any length under water. When it was first proposed to lay a cable across the British channel, it was thought a mad project, that could only end in utter failure. Even after messages were received in London, the whole thing was denounced as a hoax, an imposition on the credibility of the public.

Field pointed out to his audience that of fifty-three major cables laid, forty-five had been 'a perfect success'. He did not mention that the ones which had failed had been the longest, and deepest. Although he would never have admitted at the time, for in truth he did not believe it in 1857 and 1858, he now said that those earlier attempts to span the Atlantic had been 'a mere experiment'. He tried to entice the merchants by reminding them of the huge commercial success of the telegraph to the Pacific. Although it served essentially only one state, California, and had to pass thousands of miles of barren plains and mountains, that cable 'has paid for itself *in one year*'. The comparison, though, between the well-tried overland technology and a submarine line, was not a real one, as Field knew. With the ocean cable, although the potential traffic was much higher, the cost was vastly more and the technology by no means certain to succeed.

Field's efforts to raise money in America were disappointing. The cable, while it might appeal in theory to the provincial merchants of Britain and the United States, still appeared too risky a scheme for them to commit their own money. Field talked about the first cable as 'a leap in the dark', but he had not made that clear at the time, even if he had then believed it to be true. He assured his American audiences that much had been learned in a few short years, about instruments for testing, about cable insulation and the conductivity of copper, and about the mechanical problems involved in laying a line from a moving ship. He confessed that the work had been attempted too rapidly, and that other mistakes had been made. His optimism was undimmed, and he believed that it would, only four years later, be 'an easy task to lay a cable constructed and submerged by the light of present experience'. But still, the central problem would not go away, that the cable's success was not a certainty. An investor in the Atlantic Telegraph Co. stood to lose everything, as many had already. Any number of government guarantees on the rate of return could not get round this. It was clear that a working line would be vastly profitable, but a promise of great returns did not outweigh the fear of total loss. It was this, rather than the uncertainties of wartime, which deterred the merchants of the eastern seaboard, and in Manchester, Liverpool, Glasgow, and even London, from risking their money on a new cable.

6

A Thrill along

the Iron Nerve

There was to be no cable expedition in 1863, but events turned at last in the Atlantic company's favour. The first inkling of a real revival was late in 1862, when the cable-makers Glass & Elliot came up with an idea of how they might cover the costs.

This opened up a new way forward. The inspiration had come from Richard Atwood Glass, who had been an accountant before he became a cable-maker. There had been some timely reforms of company law in 1856 and 1862, establishing limited liability, and Glass saw how this new form of company could be an instrument to square the cable finances. In simple terms, it involved paying bills with company shares, avoiding the need for ready cash in advance. No longer would the Atlantic company have to depend entirely on individual investors – George Saward, pursuing the provincial merchants of England, had had as little luck as Cyrus Field in America – or on elusive public funding. Glass's proposal was this: his company would take payment only for materials and labour costs, until the cable was successful. The balance of the debt could then be settled in Atlantic company shares. The cable-making company would also make an investment of £25,000 into the Atlantic project. Glass & Elliot were manoeuvring hard for the valuable contract, offering 'the most liberal conditions'.

So the cable contractors were poised to take the place of those wealthy New Yorkers who during the 1850s had written cheques to cover the Atlantic company's first million dollars, but who were no longer able or willing to finance the bid. Glass & Elliot's plan meant that the contractors had a direct incentive in the venture's success. This also satisfied one of Galton's complaints about the arrangements in 1857. And who better to shoulder a substantial part of the burden? Submarine cable companies possessed the technical insight to understand just how far risks had decreased, and how lucrative the telegraph company shares might turn out to be. The cable-makers also urgently needed the Atlantic crossing to be completed, as its success would herald many more long deep-sea lines. Without a transatlantic line in place, the future of the cable industry would be limited indeed.

Soon after the expedition of 1858, Glass & Elliot began to recruit their own electricians and engineers and started to offer a full contracting service, including cable-laying. Their new staff, among whom were Bright's former assistants Samuel Canning and Henry Clifford, added to their deep-sea experience on contracts for the French and Italian governments in the Mediterranean during the early 1860s. The company was doing well, moving up to leader in the field. Yet Glass & Elliot would not take on the entire risk for the Atlantic cable, 'as we consider that would be too great a responsibility for any single firm to undertake'. They did offer 'to stake a large sum' as a measure of their confidence, and the partners also made large personal investments.

Despite the cable-makers' vote of confidence, there was still a gap in the finances. The Atlantic company went ahead anyway, in August 1863, in advertising for cable-makers to tender for the contract. Applicants were asked to submit their own specifications for the cable, and all seventeen proposals received were turned over for scrutiny by the Atlantic company's newly formed scientific committee. This consisted of Galton, Wheatstone and Thomson as well as the mechanical engineers William Fairbairn and Joseph Whitworth. To no one's surprise the successful bidder was Glass & Elliot, by a unanimous decision. The scientific committee insisted on further tests and improvements to the winning cable design, to which Glass & Elliot willingly agreed.

Field believed that the necessary funds were in place, and told New York shareholders that in October 1863, at a meeting where his energy and perseverance were again praised by his old supporter, Peter Cooper. It had taken all of five years, said Field, 'to restore the confidence of the public to such a degree as to make it ready to renew the undertaking. In business, as in war, it takes a long time to recover from the blow produced by a great disaster.' Field credited Galton with breathing new life into undersea telegraphy. 'This testimony revived public confidence in Great Britain, which was further strengthened by the success of other long submarine lines, especially that between Malta and Alexandria.' The perennial optimist went so far as to predict a direct line in place between the United States and Europe in 1865. He assumed the Atlantic telegraph's problems were over, and was already thinking of the next project.

But the marriage between 'star-eyed Science' and 'sordid Mammon', as Henry Field put it, was again delayed. The scheme stuttered early in 1864. The Atlantic company made another fruitless plea to Gladstone. Through their 'utmost exertions' they had received £320,000, from more than a thousand individuals; 'each,' it was claimed, 'being chiefly influenced by the national character of the enterprise.' The cable was designed 'by a committee of eminent and scientific men after a long series of experimental tests and after a full enquiry into the whole subject'. But a deadline drew near, as the whole capital had not been pledged, so that the money raised would have to be returned to investors early in March, and the 'great national enterprise abandoned at the very moment when it is ripe for success'. This meant no transatlantic cable for years to come, unless a national emergency demanded it, in which case the whole cost could fall to the government. But Gladstone was again unmoved.

At this point, Cyrus Field had an encounter which dramatically changed the company's prospects, comparable to his own meeting with Gisborne in New York a decade earlier which had seen the rescue of the original Newfoundland project. The network of political contacts he had cultivated in London during the civil war turned up an unexpected bonus for the Atlantic project. John Bright introduced Field to Thomas Brassey, a financier noted for his broad and imaginative investments. Brassey grilled Field on every aspect of the scheme, paying close attention to profit predictions. He then agreed to back the cable, and to recruit other capitalists to do the same. John Pender was the second to join after Brassey.

John Pender.
(*Cable & Wireless Archives, Porthcurno*)

Pender, a millionaire Manchester textile merchant and MP, already had interests in telegraph companies. Daniel Gooch, best known as chairman of the Great Western Railway, also signed up in support.

Salvation was therefore in sight as Cyrus Field marked the tenth anniversary of his first trip out of Boston on Atlantic company business. He had since crossed the Atlantic thirty-one times. The occasion was celebrated, in the words of Stuart Wortley, by 'a company of distinguished men – members of parliament, great capitalists, distinguished merchants and manufacturers, engineers and men of science, such as is rarely found together even in the highest house in this great metropolis. It was very agreeable to see an American citizen so surrounded' – gathered at Field's table at the Palace Hotel, Buckingham Gate.

The following day, a meeting of company shareholders heard details of tests carried out in Manchester on various

cable samples sent by Glass & Elliot to the scientific committee. Based on these, and on past experience, the cable's design
was decided. A seven-strand copper conductor core would
be embedded in Chatterton's compound, and insulated with
gutta percha and Chatterton's alternately, eight layers in all. It
was padded with hemp and preservative, and protected by ten
solid iron wires, each clothed in yarn and preservative. The
weight was almost double that of the 1857 cable, at 1.7 tons
per nautical mile, with a vastly improved breaking strain of
7¾ tons, more than double that of the previous line. Coiled
in one circle, it was estimated that the new cable would make
a pile 58ft wide and 60ft high. The cable was relatively small,
an inch in diameter, and required little charge. There would
be no repeat of Whitehouse's massive voltages. Henry Field,
the minister brother of Cyrus, explained that 'God was not in
the whirlwind, but in the still, small voice. A soft touch could
send a thrill along that iron nerve.'

Improvements in conductivity, insulation and strength
came at a price, one of £700,000. Another extraordinary
meeting of the company was called on the last day of March,
to authorise an increase of capital and allow fund-raising
through bonds and mortgages. The cable would be paid for
partly in shares and old debentures of the company, with
later payments in old, unguaranteed, shares. Thus the Atlantic
company itself need find no more than about half the cost of
the cable at the outset. They had £316,000, and required perhaps a further £34,000 in order to proceed, but the scheme
had been saved. For the first time, the directors dared to talk
openly about the high returns which could be expected.
'If, as we confidently anticipate, we are perfectly successful',
said Stuart Wortley, 'we shall then be in a position to earn

very large and perhaps unparalleled profits.' Others shared this faith, for company stock was rising. Shares of £1,000 which had fallen as low as £30, now stood at £350 or more. The company proposed to issue £50,000 more in guaranteed eight per cents.

Another celebration was in order. Field threw an inauguration banquet at the Palace Hotel on 15 April 1864, to commemorate the Atlantic company's renewal 'after a lapse of six years' of the endeavour to unite Ireland and Newfoundland. The guests included the American ambassador Charles Francis Adams; John Bright, MP; Galton and many of the engineers involved, including Latimer Clark, William Fairbairn, Cromwell Fleetwood Varley and Charles Wheatstone; the banker and MP Samuel Gurney; Julius Reuter; and city financiers including Brassey and Glass as well as Pender.

The new plans required a restructuring of the cable industry, with the merger of two leading companies. The coremakers, the Gutta Percha Co., amalgamated with Glass, Elliot & Co. to form the Telegraph Construction & Maintenance Co. Ltd, known as Telcon. This had become very big business, with the new company capitalised at £1 million. Its directors included Henry Ford Barclay, Thomas Brassey, Richard Glass and George Elliot, Daniel Gooch, John Pender and Samuel Gurney. Pender was chairman, and the dynamic accountant Glass, managing director.

The cable had evolved, from something a few years earlier 'regarded merely as a scientific experiment', to being 'now absolutely necessary for the purpose of our social system'. Moreover, it was promised to be 'a highly remunerative investment'. Telcon could boast unparalleled experience in the field. The Gutta Percha Co. had in the previous decade made over

9,000 miles of insulated wire for the inner cores of submarine cables. Glass & Elliot, their main customer, had made and laid more than 6,500 miles of underwater cable. The merger produced a completely integrated cable-making and laying service.

Some of Telcon's confidence and authority, not to mention the transformed financial prospects, gave new strength to the Atlantic company, which was spurred to withdraw at last from its agreement with the British government. In practical terms the contract had proved useless, doing nothing to improve the chances of success, with a guarantee which was worthwhile only if the cable made little or no surplus. Yet the promoters knew by now, from the experience of other deep-sea lines, that once a telegraph was working, it cost little to run and could expect handsome profits. And while the government guarantee was not very helpful, it carried with it some obligations, such as priority for official messages on the line. The condition causing most inconvenience for the company was the government control of telegram prices.

Matters came to a head in 1865, when the Atlantic company, expecting the cable to be working that summer, wanted to increase charges. This had to be justified to Gladstone. The directors reminded him of their past losses, totalling £465,000 in 1857 and 1858, and that they were now spending a further £835,000 which had been raised with difficulty. They were keen to 'disclaim very earnestly the slightest desire to exact for themselves or their shareholders any greedy or excessive gain'. Experience from the Mediterranean cables, though, had forced a reassessment of the likely transmission rates. It was estimated that a transatlantic line could carry only about 150 messages in twenty-four hours – 75 in each direction.

The company also considered how much business the cable might attract. The fastest way to send a message from London was by cable to Queenstown (Cóbh), south of Cork, where steam-powered British mail packets en route from Liverpool to New York were intercepted with packages of perhaps 200 telegrams at a time. As the ship passed Newfoundland, these were thrown overboard off Cape Race, from where they were wired to New York. This service was four days quicker than the post and despite a cost of £2 17s for twenty words, attracted plenty of business. As the cost of a direct message on the new telegraph was still fixed at £5, the company was concerned that without a price increase they would be swamped with as many as 500 telegrams a day, so that 'the office would soon be in the greatest confusion'. This weight of messages 'would crush the profits and destroy the usefulness of the cable', for the company had to be able to guarantee speedy transmission.

Did the contract with the government still have force? No one was quite sure. To the evident relief of both parties, the agreement was cancelled before the attempt to lay a cable in the summer of 1865. This left the way clear for the company to levy charges, and organise its business, as it thought fit, and spared Gladstone a decision he did not want to make.

New technologies and new ways of working – in engineering, electricity, finance and business – had saved the day. The final part of this jigsaw of coincidences was the most dramatic. It was the arrival on the scene of Brunel's gigantic iron steamship, the *Great Eastern*. Five times larger than anything else afloat, she was the only ship in the world that could accommodate the new cable in its entirety. As a cable-layer, the *Great Eastern* enjoyed the only successful phase of her career.

At other times she brought bankruptcy to a succession of owners, losing in total $5 million. Her conversion from passenger liner to cable ship took place in the Medway in the spring of 1864, and Captain James Anderson was persuaded to take leave from the Cunard Line and assume command. Some believed that God had sent 'the Leviathan' purposely to carry out this great work across the Atlantic.

The truth was more prosaic. It had long been noted that this ship, above all others, was ideally sized to lay the Atlantic cable. She had spent much of her short life idle and unproductive, passing through a number of owners. Daniel Gooch stepped in when she came on sale again, and bought her through the Great Eastern Steamship Co., set up with himself as chairman and Thomas Brassey among the directors. She was refitted as a cable ship for £15,700, with three vast water-filled cable tanks spread along the ship to distribute the weight. Clifford and Canning installed improved cable-laying machinery, and for the first time there was equipment to grapple and recover the cable if it broke. Gooch's plan made no concession to failure. The *Great Eastern* was to lay the cable at no charge, and only in the case of success would the owners receive payment, of £50,000 in Atlantic company shares. This was all arranged by the end of 1864, by which time Telcon had the new cable under construction and it was being stowed, and continually tested, on board the *Great Eastern* where the water tanks protected and preserved the gutta percha insulation.

In April 1865, as the American Civil War finally closed in victory for the north, Abraham Lincoln was assassinated. News of the president's killing, on 15 April, was rushed across the ocean by the steamship *Nova Scotian*, reaching a telegraph station in Londonderry on 26 April. The despatch arrived in

London two hours later and appeared in the British newspapers the following day, twelve days after the assassination. Julius Reuter, first with the news, was later accused of having personally profited on the stock market through holding back his early knowledge. The shock value of the report was somewhat muted as so much time had passed since the event. But the value of Reuter's coup gave some perspective to the proposed costs of Atlantic telegrams.

The *Great Eastern* was still swallowing 'ship-load after ship-load' of cable, 'as if she could never be satisfied'. A visitor to the ship in May, when almost all the line was on board, 'was at a loss to find it'. Stowing was finished by the end of May, although it was another six weeks before the expedition was ready to embark. There would be 500 men on board, as well as live cattle, oxen, pigs, sheep, ducks, geese and hens to provide fresh food. The projectors had taken another lesson from the *Agamemnon*'s misfortunes, and would not risk being reduced to eating three-year-old salted beef. There would be no spectators on the voyage, and even directors were permitted only if they had a specific job to do. Field and Gooch were the only representatives of the companies' boards. Thomson and Varley would be there, but with strictly defined and limited powers of interference. There were several artists and one journalist, William Howard Russell of *The Times*, who had made his reputation describing the horrors of the Crimea, to document what was expected to be an historic event.

Of all those on board, Cyrus Field was the only American. Nor did the *Great Eastern*'s escort contain any United States vessel, for Lincoln had declined to help, in view of the continuing bad relations between his victorious northern states and the Palmerston government. The expedition was to be accompanied only by two British warships, *Terrible* and *Sphinx*. The *Caroline* and *Porcupine* were also in assistance. Canning was engineer on the *Great Eastern*, assisted by Clifford; chief electrician was Charles de Sauty, who had been involved almost from the start in the Newfoundland lines and was also the man who had struggled alone in the telegraph station there as the 1858 cable died.

Because of the great size of the cable-laying ship, the landing points at both sides of the ocean had to change. Instead of Valentia harbour, the cable would leave from Foilhommerum Bay five miles away, overlooked by a tower built by Cromwell after the English Civil War. In Newfoundland, Field had found an alternative landing, a deep but sheltered spot near the open sea called Heart's Content. The shore end laid, the *Great Eastern* steamed away west from Ireland on Sunday, 23 July 1865.

Only a few hours later, with eighty-four miles laid, a fault was discovered, though the line still worked. The ship retraced her course, picking up the cable, for more than ten miles. There, a minute piece of iron wire, smaller than a needle, was discovered to have pierced the cable, destroying a small section of insulation. It was repaired, and the journey continued.

All continued well until the following Saturday, with 800 miles laid, when the line went dead. The hauling-in equipment had its first real test, in two miles of water, but without difficulty brought in the damaged section for repair. Much to Canning's alarm, there was no doubt that an act of sabotage had taken place, for again a piece of wire had been driven through the insulation. The first fault might have been an accident, but this clearly was not. Something similar had happened before, on a North Sea expedition when a rival company was supposed to have bribed a workman to destroy the cable. On the *Great Eastern* it proved impossible to identify the culprit, and all that could be done was to place a watch on the cable men.

After this, spirits lifted, for everything went well, the machinery and systems running smoothly. By Wednesday, ten days into the voyage, 1,200 miles of cable were laid and the ship was 600 miles from Newfoundland, two days away from shallow coastal waters. From Ireland, where the signals received had been so good that it was possible to detect when the ship rolled, news of the expedition's progress was reported daily to the London papers. But suddenly there was a complete loss of communication.

The cable-layers had noticed a small fault, which was overboard before it could be mended. Though it was a tiny one, a pin's prick, it needed repair. There was no great alarm, and the crew prepared to haul in the cable again. As steam was got up on the small engine used for this, the ship drifted across the line and chafed it. When it was pulled over the side, this damaged section snapped, and the end of the cable disappeared into the ocean.

Canning decided that there was only one thing to be done, to grapple for the line in the 2½ mile depths. He had been noted for impressive cable-fishing exploits in the Mediterranean, but this was much more ambitious. The grapnels, two five-armed anchors ending in hooks, took two hours to sink to the ocean bed. The *Great Eastern* then had to trace a new route a few miles from the broken end, across the line where the cable had been laid. She sailed back and forth all night, trying to snag the tiny line so far below. At last there was success, and the grappling irons caught the line. Canning knew that it was not just a stray object, for the strain increased as the load was pulled nearer the surface.

For hours they worked to raise the cable. When it had been lifted three quarters of a mile from the ocean bed, an iron swivel gave way and the catch, along with two miles of rope, returned to the bottom. After a delay with fog, they tried again, and again. Three times the cable was found, brought up part of the way, and lost again as the equipment failed. Every time, more of the wire rope was lost, until, after the fourth failure, there was no rope left, and no alternative but to mark the spot with buoys and abandon the effort for the year.

This was a strange kind of failure, for it had 'the moral effect of a victory'. The returning ship and crew were greeted as heroes. The great ship had proved her value on such an expedition, safe and manoeuvrable in any weather. The grappling attempts had nearly succeeded. The laying machinery had been a stunning success, and hauling in had worked well. Faults had been spotted and located with impressive speed and accuracy. Best of all, though, had been the ease and clarity of sending messages along 1,200 miles of cable at the bottom of the ocean. It was now clear that eight words a minute, at the very least, were achievable.

By the time the *Great Eastern* arrived back in Ireland in the middle of August, Cyrus Field, with Canning, Glass and Thomson, had prepared a prospectus for an attempt the following year. Field would not again need to dust down his well-worn arguments to keep the project alive, for it was clear that triumph was not far away. In fact the American had little trouble in convincing his fellow directors that the 1866 expedition should pick up and complete the abandoned 1865 cable, as well as lay a new line.

Telcon was already making a new cable in September 1865, when the Atlantic company announced the share offer which would pay for it. There were to be 120,000 new £5 shares, with the promise of a twelve per cent dividend with priority over all other investors. Telcon would receive £500,000, and a bonus of £100,000 conditional on success. The shares were quickly taken up, although it later emerged that half had been sold to Telcon itself.

The engineers were confident but far from complacent. Work to improve their instruments and systems continued through that winter. William Thomson refined his theoretical work, presenting a paper to the Royal Society of Edinburgh in December, 'On forces concerned in the laying and lifting of deep-sea cables'. Willoughby Smith, formerly of the Gutta Percha Co. and now an indispensable figure in the project, devised a new method for the electrical testing of cables during submersion, which Thomson approved. This meant continuous testing of the cable insulation, and instant detection of any fault. The cable design had been again improved, the armour lighter and stronger, with galvanised iron wires to protect from corrosion so that a smaller coating of compound was needed. Grappling and hauling machinery was further refined. The company was also looking for better ways to run the line, casting around for more sophisticated telegraph codes. Frank Bolton of Chatham, racing to complete one such code, had tried an unfinished version through 2,300 miles of cable on the *Great Eastern* in 1865, claiming to save almost half the time of previous methods.

In spite of all the planning and preparation, the Atlantic Telegraph Co. was still to suffer another crushing and unexpected blow. On his return to England on Christmas Eve 1865,

Field was told that the Attorney General would not allow the twelve per cent preference share issue to proceed. It was illegal under the company's own Act of Parliament. The only way to save the preference share scheme was to amend the Act, and this was impossible to achieve in the current parliamentary session. In desperation, the company again appealed to Gladstone. Could the government subscribe the quarter of a million pounds capital which private investors would not commit without the high returns of a preference scheme? The cable's prospects had been transformed since previous requests to the Chancellor, but this final appeal was equally fruitless. What prospect was there now that money could be raised in time for the short cable-laying season of 1866?

Although the outlook was bleak, the projectors decided to carry on making the cable in the hope that something could be done in the short time available. Field remained upbeat, telling the directors of the New York and Newfoundland company early in March 1866 that he had little doubt of success that year. And finally, a way was found to circumvent the legal problems surrounding the Atlantic company. Field took advice from Daniel Gooch, who, despite his close connections as a director of Telcon and of the Great Eastern company, had remained aloof, never personally investing in the Atlantic company nor completely convinced by its prospects. But he had changed his mind after witnessing at first hand the events of 1865. Gooch suggested forming a new company, and this time offered to invest substantially himself.

As a result, the Anglo-American Telegraph Co. Ltd was founded. It undertook to lay a cable for the Atlantic company that summer. In return the Anglo-American was entitled to a huge slice of future income, with priority over claims by Atlantic shareholders or creditors. This was later to cause untold difficulties. Atlantic company investors who had kept faith over many years and suffered great losses, now found themselves cut out of the profits. Furthermore, Thomson and his engineering associates, who had given time and expertise unpaid in the expectation that they would reap benefits later, were ignored under the new arrangements. The Anglo-American, it emerged, did not consider that anything was owed, legally or morally, to these men, although the cable could never have been laid or worked without them. The directors, who included Glass and Gooch, knew this full well. They had even relied on the engineers to help float the new company, asking Thomson, Jenkin, Varley and others to write assurances in the prospectus that a rate of eight words a minute down the line could easily be achieved.

Once again, there had been a last ditch rescue. But Field had again been over-optimistic about finance. Very little money was subscribed by the general public, who, lacking the confidence of the insiders, still feared losing all. A total of £230,000 had been promised by private investors, mainly people connected with Telcon. Telcon itself pledged £100,000. Ten individuals, among them Elliot, Glass, Gooch, Pender and Field himself, were committed to investing £10,000, but there was still a shortfall. This time the cable was saved by the merchant banker Junius Morgan, now principal of J.S. Morgan & Co., the friend of Field from the days of pro-Union lobbying in London. Morgan's bank committed itself in April 1866 to supply the rest of the capital Field needed. By the utmost good fortune, this agreement was sealed just before 10 May, Black Friday. The fall-out from this financial crash would otherwise have ended any chances of raising money that year, and maybe for much longer. And by strange irony, Black Friday was triggered by the collapse of the bank of Samuel Gurney, long-time director of the Atlantic company.

But it was a done deal, and there was to be no other significant obstacle to the cable. Success when it came was an anticlimax. The *Great Eastern* was accompanied by the *Terrible*, the *Albany* and the *Medway*, with *William Corry* laying the shore end at Valentia. Anderson was again the master, Canning the engineer, Willoughby Smith had taken over as electrician, with Thomson on board lending advice. The flotilla set sail on Friday, 13 July 1866. It was, said Henry Field, 'almost monotonous from its uninterrupted success'.

As planned, the ship proceeded slowly, covering between 105 and 128 miles a day. The British public had daily newspaper accounts of progress as the cable unrolled across the ocean floor. Across the Atlantic, there was no news – without a cable, nothing could yet be known. In Newfoundland, some people, though not as many as the previous year, waited anxiously, expectantly, at Heart's Content for news. At daybreak on 27 July, as fog lifted, there was wild excitement, 'the wildest excitement I have ever witnessed,' said Gooch, when masts were spotted on the horizon. There were six of them, and four funnels, and it was without doubt the *Great Eastern*. A fleet of small boats set out to meet the successful convoy. The submarine telegraph between Newfoundland and Cape Breton had broken down, and so it took two days for the tidings to reach New York. The first news Field's instrument of peace brought to the New World was of war, between Austria and Italy in the Adriatic, but the report ended with confirmation of a peace treaty. This was taken as an excellent sign, that the cable 'was born to be the herald of peace'.

Although the British government had been at times frus-
tratingly unhelpful – and the Liberals were by then out of
office – one of the first telegrams to cross the Atlantic on
the new cable was sent by Field to Gladstone. 'Many thanks
for your kind words last night in the House of Commons
about my country which I read with much pleasure here this
morning soon after breakfast – very truly your friend, Cyrus
W. Field.' Written from Heart's Content, here was the
announcement, if one were needed, writ large, of an overnight
revolution in transatlantic news and communication.

The Cabot Strait cable was repaired in two days, and the
line fully opened for a predictable rush of congratulatory
messages. Even better news was to come. By the time *Great
Eastern* had refuelled and made her way 600 miles back east,
Albany and *Terrible* had located the 1865 cable and almost
caught it. It took days longer, and thirty frustrating attempts,
to secure and retrieve the line. Once on board it was carefully
tested. It worked. Willoughby Smith could communicate with
Valentia, where two telegraphers who had spent a year testing
the defunct cable were startled to receive coherent signals at
last. For the first time, Cyrus Field wept.

A triumphant message to Field's family in New York went
from the ship to Valentia, and back across the ocean on the
new cable. Then the *Great Eastern* steamed slowly to St John's
through bad weather, completing the 1865 line on 7 September.
The second round of celebration was even more exuberant, for
the revival of the lost cable was to the public more exciting still
than the laying of the new.

Field was lauded and honoured, and later given a banquet
by the New York Chamber of Commerce at the Metropolitan
Hotel, attended by the acting vice president of the United

States, Lafayette S. Foster. In London, *The Times* joined in the general hyperbole. 'Since the discovery of Columbus nothing has been done in any degree comparable to the vast enlargement which has thus been given to the sphere of human activity.' Captain Anderson, Professor Thomson, Canning and Glass all received knighthoods; Gooch and Curtis Lampson, directors of the Anglo-American Telegraph Company, the higher rank of a baronetcy. Field, as an American citizen, could not be knighted but was heaped with praise by the British establishment, and awarded a Congressional gold medal in the United States.

The delight was not quite universal. In Siberia, gangs of workmen erecting 20,000 poles for the Russian-American overland route threw down their tools and walked away when news of the Atlantic cable belatedly reached them. Western Union lost $3 million on that venture. Elsewhere there was some disapproval of the Revd Henry Field's triumphalist account, and the way that he invested the cable with religious meaning. A critic in the American *Southern Review* pointed out that the cable was in fact 'about as [religious] as Wheeler's Sewing Machine'. The Atlantic telegraph was:

> simply a postal arrangement. The utmost that can be effected by it, is the transformation of intelligence between Europe and America eight or nine days earlier than before. This is a matter of importance. It facilitates commerce and the capture of absconding criminals, it serves travellers and will be of great comfort to many an anxious heart. We can also imagine instances in which great international interests might be secured, which the interposition of some days might put in peril. These are advantages to rejoice in, and

be thankful for … But let the praise be discriminating, and then it will be at once more sincere and more valuable.

The critic also thought Americans were trying to take more than their share of credit for the success. 'One might suppose, from the style and tone of the demonstrations in New York, that it was as thoroughly American as if it had been the out-growth of the Monroe doctrine.' In fact, the United States' involvement extended only to the fact that Field was an American citizen, and that some of the initial surveys had been carried out by the US Navy. 'For the rest, it was done chiefly by British science and mechanical skill, British enterprise and British capital.'

This, though, is hardly a full story. The cable had been driven forward from the start by Americans, from the Canadian provinces and the United States. Britons had brought to it a sophistication in science, technology and finance. The honours were similarly divided between engineers and entrepreneurs, between the ideals of international peace and understanding and the desire for commercial profit.

The new telegraph was so expensive that very few could afford to use it. As Sir James Anderson put it, 'no one can order his dinner by telegram'. Traders, the cable's main users, drew great benefits from having the latest and most reliable commercial data. The effect was to stabilise financial markets, especially in the United States, where solid information replaced damaging rumour and speculation. And the cable was itself, after so many tribulations, a major business success. In its first year, the Anglo-American Telegraph Company paid a dividend of more than twenty-five per cent. Telcon benefited too, John Pender reporting to shareholders in 1867 that trade was booming, the company's competence to make, lay and recover cables 'in great oceanic depths' settled beyond doubt.

But Gisborne's 'mystic voice of electricity' had a wider impact yet on millions who could never pay to use it directly. On the eve of the first attempt to lay a cable in 1857, the Lord Lieutenant of Ireland had predicted that the cable would give connections between the Old and New Worlds 'a life and an intensity which they never had before'. The cable's part in bringing international peace is questionable – better communication does not necessarily bring better understanding. In other ways, though, this life and intensity was immediately apparent. Reuter was the first to capitalise upon it, his agency's telegrams packed with the latest news for the following day's papers across the ocean. When James Garfield was shot in 1881, reports appeared in British newspapers within hours, a stark contrast to the twelve days taken over the previous presidential assassination. From this moment began a sense of shared experience, a convergence of cultures, between the two English-speaking nations.

So dependent was Reuter on the transatlantic cable – and so increasingly irritated by the Anglo-American's high charges for telegrams – that in 1869 he was instrumental in launching the first direct line from Europe to the United States, between Brittany and Boston. Sir James Anderson was also involved, as director of the expedition, and he subsequently became a successful businessman, director of a number of submarine telegraph companies associated with John Pender. He rose to be managing director of the Eastern Telegraph Co., an amalgamation of four of Pender's companies.

Pender himself, like Cyrus Field, had flirted with personal ruin by gambling almost everything on the Atlantic cable. He went on to found a string of submarine cable companies until he, Gooch and other associates controlled telegraphs stretching from Britain across the Far East and Australasia. Porthcurno, an isolated sandy cove in the west of Cornwall, was for a century the point at which Pender's network of fourteen telegraph systems converged, the main communications gateway of the British Empire. Pender was angered not to have been knighted in the wake of the success in 1866, nor afterwards for his services to Empire, and wrote confidentially to Gladstone in 1881 to stake his claim to a baronetcy. He argued that he had risked a quarter of a million pounds on the Atlantic scheme, 'without which the necessary capital for the undertaking could not have been secured'. His entire motivation, he said, was public service, and somewhat surprisingly claimed that 'pecuniary advantage to myself was neither the object nor the result of these transactions'. Pender made a further fortune through his network of telegraphs, absorbing them into the group later named Cable & Wireless Ltd. A knighthood finally came in 1888.

Canning, Bright, Jenkin, Thomson, Willoughby Smith and Varley carved out distinguished careers in engineering. Only after the threat of legal action, and mediation by Richard Glass, did Thomson, Jenkin and Varley start to receive their dues for their work on the Atlantic project. Licence payments from the company for use of their instruments were to prove immensely profitable over the following years.

Others involved in the cable expeditions of the 1850s were less fortunate. Maury, in London for several years after the Civil War, eked out a living writing geography text books. His direct association with the Atlantic cable was over, but he did receive, among other honours, an LLD from the University of Cambridge in recognition of his achievements. In 1868 he returned to the United States as Professor of Meteorology at the Virginia Military Institute in Lexington, where he died in 1873. Galton's damning conclusions ended Wildman Whitehouse's career in telegraphy in 1861, and he spent the rest of his life experimenting on electrical devices for trams and omnibuses rather than returning to surgery. Frederic Gisborne, having fallen out with Field at the end of 1856, abandoned telegraphy in disgust and took up mining exploration in Newfoundland and neighbouring provinces. He resumed his daring and strenuous life of old. A serious gunshot wound during one of these expeditions in 1861 forced him to find a less active role. Gisborne made a great deal of money as a minerals agent, but lost it all in the 1870s. The Canadian federal government then appointed him superintendent of the Dominion Telegraph & Signal Service, and for the remainder of his life he criss-crossed the remotest parts of the country, building and re-building telegraphs. At the time of his death in 1892, he was engaged in plans for a Pacific cable.

For the 'changed world' which *The Times* had foreseen in 1856, the unceasing march of progress up to the century's turn, had still not delivered the Pacific telegraph. This was one thing that Cyrus Field had not been able to effect, although he too tried. Having staked most of his assets on the Atlantic cable, Field was afterwards able to pay off all his remaining creditors from 1860. He never gave up on his mission to improve relations between the United States and Britain, acting as a conduit for information between John Bright and colleagues, and their American equivalents. But this alone did not satisfy him, for he could not live without some grand project to fill the void left by the Atlantic scheme. For a time, the Pacific project took over. In 1880 Field was writing to an acquaintance, 'I hope that you and I may live to see a cable between the west coast of the United States and the Sandwich Islands, Japan and Australia'. He was actively pursuing it through government and Wall Street contacts, looking for 'fifteen to twenty millions of dollars, cash'. It could be complete in two years ...

By the time Field died in 1892, he had long-standing financial troubles, much of his fortune having been wiped out in saving the New York elevated railway system from collapse. The Pacific cable idea was long abandoned, and it was to be another decade after his death before it was completed. A generation passed, thirty-six years, between the first Atlantic and Pacific cables. The delay was due partly to the much greater depths and distance of the Pacific, but that is not the full story. Once the Atlantic had been crossed, a Pacific link was, at least for a time, superfluous. With Pender's cables laid to Australasia, the whole world was in communication by way of the Atlantic, as Field had predicted to the Boston merchants. Lines from Europe to the west turned out to be much busier and more profitable than the Far Eastern telegraphs, where there was much less commercial traffic. The delay in crossing the Pacific therefore further underlines the success of the Atlantic cables.

So whose idea was the Atlantic cable, and whose accomplishment? The honours are not easily awarded. Samuel Morse was the first to work out practicalities, and he continued to support the project, though his authority as a leading electrician had evaporated some time before the line was complete. Frederic Gisborne's energy and insight had given impetus to the scheme, although he too was sidelined as engineer, even before the first attempt on the Atlantic in 1857. Gisborne's most significant feat, as it turned out, was to bring in Cyrus Field. Field had no equal in driving forward the Atlantic telegraph. Without him, the wild and visionary scheme might not have been attempted until years, maybe decades, later. Were it not for Field's over-ambition during the 1850s, there would have been no failures then, no Galton enquiry and consequently no hard lessons about the behaviour of electricity in long cables. Out of failure came the foundations for a monumental achievement in the 1860s.

Cyrus Field was inspired, suffering endless frustrations when he could not inspire others to the same degree of enthusiasm and commitment. But he pursued the dream doggedly, even when it seemed doomed. Field's essential gift, apart from his persistence, was an ability to find the very best, the most useful, talent on offer. The network he built embraced engineers and scientists, financiers and merchants, naval officers and politicians, British and American. The result was the extraordinary story of the Atlantic cable, a feat outside its time.

Sources &

Information

The Cable is written for non-specialist readers, so does not have footnotes or a bibliography. The story is told elsewhere with full references:

Cookson, Gillian, "'Ruinous competition": the French Atlantic Telegraph of 1869', *Entreprises et Histoire*, 23 (1999), pp. 1–16
Cookson, Gillian, 'The Golden Age of Electricity', in I. Inkster (ed.), *The Golden Age: Essays in British Social and Economic History, 1850–1870* (Ashgate, 2000), pp. 75–86
Cookson, Gillian, & Colin A. Hempstead, *A Victorian Scientist and Engineer: Fleeming Jenkin and the Birth of Electrical Engineering* (Ashgate, 2000)

The main archival sources used are located: in the Institution of Engineering and Technology library and archives; in the Cable and Wireless Archive at Porthcurno; at the New York Public Library, Wheeler Collection; and within the Kelvin papers in the university libraries of Glasgow and Cambridge.

Dibner, B., *The Atlantic Cable* (1959) (http://www.sil.si.edu/ digitalcollections/hst/atlantic-cable/)
Finn, B.S., *Submarine Telegraphy: the Grand Victorian Technology* (1973)
History of the Atlantic Cable website (http://www.atlantic-cable.com/)
Porthcurno Telegraph Museum, Cornwall
Russell, W.H., *The Atlantic Telegraph* (1865, reprinted 2005)

Index